SECRET SCIENCE AND THE SECRET SPACE PROGRAM

HERBERT G. DORSEY III

The opinions expressed in this manuscript are solely the opinions of the author and do not represent the opinions or thoughts of the publisher. The author has represented and warranted full ownership and/or legal right to publish all the materials in this book.

Secret Science and the Secret Space Program
All Rights Reserved.
Copyright © 2015 Herbert G. Dorsey III
v2.0

Cover Photo © 2015 thinkstockphotos.com. All rights reserved - used with permission.

This book may not be reproduced, transmitted, or stored in whole or in part by any means, including graphic, electronic, or mechanical without the express written consent of the publisher except in the case of brief quotations embodied in critical articles and reviews.

Herbert G. Dorsey III Publishing

ISBN: 978-0-578-15238-7

PRINTED IN THE UNITED STATES OF AMERICA

Foreword

Writing a book about secret programs that are so secret that even most in our official government are unaware of, has been quite challenging. But, I have several things going for me that definitely have been a help. I have a broad education in Electrical Engineering and Physics that go beyond what is taught at the university level. I have spoken with people that worked on "black projects" whose identities will remain secret for their own protection. And, I have personally investigated free energy inventions, which, I know for a fact, work. I have seen them working. I have also seen how these inventions, in the past, have been vigorously suppressed. Whether they will continue to be suppressed in the future remains to be seen.

I am intensely curious and a voracious reader of subjects that I am curious about. If a group of people are keeping something secret, I am highly motivated to discover what they are keeping secret and why.

That being said, I also understand that some secrets are necessary, like secret weapons technology that could fall into the wrong hands. I respect those kinds of secrets. On the other hand, secrets that could benefit mankind, that are being withheld to give an economic advantage to a greedy few, or secrets that control groups keep in order to subjugate the masses, I will gladly reveal. Some secrets I already have revealed in my previous book: *The Secret History of the New World Order*.

Much information has come from "whistle blowers" from within secret government agencies. I have met some pretty interesting people in my travels and in the UFO and free energy conferences that I have attended and have heard a lot of mind boggling information.

The problem becomes one of determining the true whistle blower from the "false flag" whistle blower and the true information from the false. One technique that I use is determining the scientific probability of real technology as opposed to unreal technology. There are some phonies out there, so one has to keep an open mind with a healthy amount of skepticism. I have tried this approach for many years and pretty much have been able to ferret out the real from the unreal.

Another technique is determining if the subject matter correlates with other information from other sources. This technique has to be used in cases where back engineered extraterrestrial technology is involved that use science still unknown to most of us.

That being said, there is most definitely a secret science used in "black projects" that is not taught at the university level - some of which incorporates back engineered extraterrestrial technology. I don't intend to go into complete detail on all the subjects touched upon herein. To do so would require many volumes. I will however, try to give references to source material for those who wish to investigate further.

Some may find a problem with the fact that there is an extraterrestrial presence on our planet. While it is true that that fact has been a highly classified government secret, it is also true that there is an overwhelming body of evidence in the public domain that we are being visited by people not of this world. Also, there seems to be a program to gradually reveal the ET presence to the public by some people in these secret programs. Only those that wish to keep their heads "buried in the sand" to preserve their "mental comfort zone" can overlook this fact.

I have trusted friends that have seen flying saucers, I have seen flying saucers. Thousands of people have seen flying saucers. There is plenty of photographic and video evidence of flying saucers. Who is flying these saucers? That question will be answered in this book. In any case, for those of you that still are not aware of the extraterrestrial presence on our planet, I will provide many sources of that information that I consider reliable.

As far as a secret space program goes. Do you really believe that the U.S. government would land men on the Moon in the 1970s and then never go any further? If so, you really don't understand how the secret part of the U.S. government really works.

Much of the technology of today would seem like science fiction a century ago. Some of the secret "black technology" is probably a century ahead of "open" technology.

You may choose to believe that this is a science fiction story, if it is too far beyond your reality level. Or you can choose to believe that it may actually be true and be motivated to investigate further. In either case – enjoy!

Table of Contents

Foreword i

1. Secret Science 1
2. The Science and Technology of Free Energy 33
3. The Physics of Time 48
4. German Flying Saucers 62
5. The Extraterrestrial Presence 80
6. Exopolitics 116
7. Modern Antigravity Development 125
8. The *Solar Warden* Fleet 155
9. Beyond Anti-Gravity 170
10. Conclusions 195

Biblography 202

1

Secret Science

The Science taught in schools and universities today is but a weak version of what is actually possible and much has been purposely omitted for economic advantage and military secrecy reasons. The following chapter will cover some, but not all, of these omissions.

Some inventors have developed antigravity as long ago as the late 1800s. John Worrell Keely had developed the science of Sympathetic Vibratory Physics (SVP). In 1872, Keely met secretly with a group of bankers. The meeting was arraigned by a leading patent attorney, Charles Collier, who had previously seen numerous demonstrations of Keely's inventions.

Keely started the meeting by stating:

> "In considering the operation of my engine, you must discard all thoughts of engines that are operated upon the principle of pressure and exhaustion, by the expansion of steam or gas. My system is based and founded on 'sympathetic vibration.'"

The scientists that Keely brought with him were of high standing and of unimpeachable reputation and they corroborated Keely's statements. They had seen him develop 20,000 to 32,000 pounds per square inch in the first experimental stages of his "Aether smashing small machine."

This meeting resulted Keely receiving loans to continue with his work.

In 1888, he demonstrated a machine using SVP that would dissolve quartz in the Catskill Mountains to 12 wealthy investors in gold mines.

His Vibratory Disintegrator was able to cut a tunnel in solid quartz 18 feet long and 4.5 feet in diameter in 18 minutes as the quartz was reduced to dust. Keely was well paid for this demonstration of his Vibratory Disintegrator. The investors kept this technology secret so that they would keep a competitive advantage in their California gold mining operations.

Keely claimed that he could dissolve atoms as easily as molecules into Aether using his SVP. He developed tremendous pressures in enclosed containers containing gases or liquids that were thusly disintegrated. Pressures as high as 25,000 pounds per square inch, were developed. He developed engines and cannons that used these high pressure forces which were demonstrated to the military and industrialists.

Later Keely announced that he had succeeded in overcoming the force of gravity and demonstrated evidence of this claim before a special committee. The event was described in the *Philadelphia Evening Telegraph* on April 13, 1890. The report stated that Keely used a model of an airship that weighed about eight pounds to which a loose wire of silver and platinum was attached. The other end of the wire was attached to a

"Sympathetic Transmitter."

When the Sympathetic Transmitter was activated the model rose up into the air. The model could be controlled to rise, descend or float stationary in the air.

After this demonstration, the bankers wanted Keely to organize a Keely Motor Company so that they could receive stock. 20,000 shares were issued, of which Keely received one seventh part.

An important concept was the "Neutral Center", which Keely defined as that central area of a sphere which contains one third of its volume. If you laid out a circle containing three smaller circles which would just touch each other and the circumference of the outer circle, another circle, whose circumference passed through the center of each of the three inner circles, would if rotated into three dimensions, have one third the volume of the outer circle if it was rotated into three dimensions. This diagram is important since it diagrams the triad of creation. Keely's molecules contained the three inner atoms, one positive one negative and the third neutral.

Although Keely's science was quite different than modern science, his ideas produced working devices. His triad of creation concept, in one way, was fractal in nature. The Molecule contained 3 inner atoms; each atom contained 3 inner particles, each of which contained 3 smaller entities and so on to ever smaller entities.

It must be remembered that at the time that Keeley was developing his concepts of molecules and atoms, very little was known about their actual structure. Although not in conformity with modern concepts of the atomic structure, it is to be noted that the nucleons of the atom are considered to each be comprised of three quarks. It would be interesting if in the future, it was discovered that each quark was also each comprised of three smaller entities. Then, it could be that Keeley was correct about his model of the structure of matter, except it didn't start at the molecular level but at the nucleonic level. Only the future will tell.

His terminology, which he had to construct because no terms existed for what he was trying to describe, was also different: "Atomoles" were units of energy, which in resonate oscillation produced the creative force called "Atomolity" which in the transmissive form created "Gravism" which transmitted through more rarified media, produces a static effect upon all other Atomoles, denominated gravity. Oscillations were also three fold; enharmonic, harmonic and dominate.

It must be stressed that Keeley's theories were based on years of experiments that he actually conducted – not on purely mathematical constructs, as so many theories today are created.

Keely was known to have moved heavy machines and weights around his laboratory using his understanding of gravity and levitation. At one point, he laid out the plans for an antigravity space ship:

> "A small instrument, having three gyroscopes as a principal part of its construction, is used to demonstrate the facts of aerial navigation. These gyroscopes are attached to a heavy, inert mass of metal weighing about a ton.
>
> The other part of the apparatus consists of tubes, enclosed in as small a space as possible, being clustered in a circle. These tubes, represents certain chords, which were coincident to the streams of force acting on the

planet, focalizing and de-focalizing upon its neutral center.

The action upon the molecular structure of the mass lifted was based upon the fact that each molecule in the mass possessed a north and south pole, – more strictly speaking a positive and negative pole, - situated through the center, formed by the three atoms which compose it.

No matter which way the mass of metal is turned, the poles of the molecules point undeviatingly to the polar center of the earth, acting almost exactly as the dip-needle when uninfluenced by extraneous conditions, electrical and otherwise.

The rotation of the discs of the gyroscopes produce an action upon the molecules of the mass to be lifted, reversing their poles, causing repulsion from the earth in the same way as like poles of a magnet repel each other. This repulsion can be diminished or increased according as the mechanical conditions are operated.

By operating the three discs, starting them at full speed, then touching two of them, so as to bring them, according to the tone they represented by their rotation, to a certain vibratory ratio, the weight slowly sways from side to side, leaves the floor, rising several feet into the air, remaining in that position, and as the disks gradually decrease their speed of rotation, the weight sinks to the floor, settling down as lightly as thistle-down. Where one molecule can be lifted, there need be no limit as to the number in a structure that may be operated upon as easily as one.

The Vessel in contemplation, the aerial navigator, will be over two hundred feet long, over sixty feet in diameter, tapering at both ends to a point, made of polished steel, and will be capable of being driven under the power of dipolar repulsion at the rate of three hundred miles per hour. It can be far more easily controlled than any instrument now in use for any phase of transit.

Another very remarkable feature connected with this system of aerial navigation, is that the vessel is not buoyed up or floated in the air through the medium of the air, so that if there were no atmosphere it would float just as readily; hence, under mechanical conditions most certainly capable of production, involving massive strength of resistance to inter-stellar vacuity this can be made capable of navigating even the remote depth of space, positions between the planets where polarity changes being controlled by other adjuncts of concentration for that purpose..." (1)

Mystery airships or phantom airships are a class of unidentified flying objects best known from a series of newspaper reports originating in the western United States and spreading east during 1896 and 1897. According to researcher Jerome Clark, airship reports were made worldwide from the 1880s to 1890s. Typical airship reports involved unidentified lights, but more detailed accounts reported ships comparable to a dirigible. Reports of the alleged crewmen and pilots usually described them as human looking, although sometimes the crew claimed to be from Mars. It was popularly believed that the mystery airships were the product of some genius inventor not ready to make knowledge of his creation public. For example, Thomas Edison was so widely speculated to be the mind behind the alleged airships that in 1897 he "was forced to issue a strongly worded statement" denying his responsibility.

In 2009, American Author J. Allan Danelek wrote a book entitled *The Great Airship of 1897* in which he makes the case that the mystery airship was the work of an unknown individual, possibly funded by a wealthy investor from San Francisco, to build an airship prototype as a test vehicle for a later series of larger, passenger carrying airships. In the work, Danelek not only lays out a plausible scenario, but demonstrates how the craft might have been built using materials and technologies available in 1896 (including speculative line drawings and technical details).

The ship, Danelek proposes, was built in secret to safeguard from patent infringement as well as to protect investors in case of failure. Noting that the flights were initially seen over California and only later over the Midwest, he speculates that the inventor was making a series of short test flights, moving from west to east and following the main rail lines for logistical support, and that it was these experimental flights that formed the basis for many – though not all – of the newspaper accounts from the era.

Danelek also notes that the reports ended abruptly in mid-April 1897, suggesting that the craft may have met with disaster, effectively ending the venture and permitting the sightings to fall into the realm of mythology. While highly conjectural, the book does make a reasonable case for the craft having been of terrestrial rather than extraterrestrial origin.

In 1969, in the University of St. Thomas, in Houston, Texas was an exhibit titled "The Sky is the Limit" which displayed 2 scrapbooks by a German named C.A.A. Dellschau (1830 -1923), who was a member of the secret Aero Club. The scrap books were filled with newspaper clippings and cryptic writings which after an interpretation by P.G. Navarro revealed the following:

In Sonora, California a secret organization called the Aero Club existed. In 1858 it was headed by George Newell. However, the club was overseen by an even more secret organization with the Initials NYMZA. This club had a number of aircraft which had dirigible shaped gas bags overhead the passenger quarters. However the volume of these gas bags using hydrogen gas would never have been sufficient to lift the weight of the ships. It was claimed that a special type of "NB gas" was used that possessed antigravity properties. Perhaps these were the craft being seen while they were on test flights across the nation in the late 1890s.

One can only wonder if Keely's technology was being used by those persons involved with the great air ship mystery of the 1890s. From all the evidence of these air ships, which were not "lighter than air craft" considering their dimensions and weight, they were operated by human beings of this planet and were very likely built by a secret organization not willing to share its secrets.

Varied descriptions of air ships were published in the Chicago Tribune, the New York Sun, the New York Herald and the New York Times. The New York Times published 3 photos of the air ship.

Newspapers of the era also mentioned John W. Keely's airship which he built and kept on developing between 1888 and 1893. His air ship successfully flew and was acknowledged by the U.S. War Department in 1896. One cannot help but wonder what ever became of that invention.

At this time, Keely Motor Stock was selling for 600 times their original value. Speculators cornered the stock and the value skyrocketed even more. In the middle of this speculative investor frenzy, Keely died. Because of Keely's secrecy, the secret principles of his technology died with him. It is possible however that certain secret societies or groups may have discovered or shared Keely's secrets and are keeping them to themselves.

About the same time Keely was amazing the public and potential investors with his SVP technology, another scientist was amazing the world with his own discoveries and inventions.

Nicola Tesla had teamed up with Westinghouse to create the alternating current system of electrical power generation and delivery. Alternating current was much superior to Edison's direct current system because it could be transformed between high voltage and lower current, requiring smaller wires to conduct the energy long distances, and low voltage and higher current required by the customers. A hydroelectric generating plant was built at Niagara Falls which supplied electricity to a large area.

From the profits of this venture, Tesla created a laboratory in which many more inventions were perfected and patented. Tesla was the primary innovator of radio communication. Marconi stole many of Tesla's ideas in his patents and developed an inter-continental wireless system. But, later Tesla won a patent infringement case against Marconi and is now recognized as the father of radio.

At Colorado Springs, Colorado, Tesla set up a laboratory to experiment with a wireless system of transmitting electrical power. His system did not use the transverse electromagnetic waves discovered by Hertz, but rather longitudinal waves transmitted from a spherical antenna which acted "like sound waves in the

Aether". He was successful in wireless transmitting useful electrical energy to a receiver 22 miles distant from the transmitter, which lit up a number of 100 watt light bulbs.

One useful discovery was that the whole planet acted as a conductor, the atmosphere an insulator and the ionosphere another conductor, the whole planetary system acting as a spherical capacitor. Tesla was able to set up standing waves inside this spherical capacitor, which circled the entire globe from his Colorado Springs laboratory. (2)

Tesla, returned to New York after his Colorado Springs experiments with the goal of constructing a wireless power transmitting station called Wardenclyffe near Shoreham, Long Island. At first, J.P. Morgan provided the financing and the construction got under way. However cost over runs caused the money to run out before completion. Tesla returned to Morgan for an additional loan to complete the project, explaining that the transmitter not only would transmit information but also electrical power. Morgan, already heavily invested in existing power line systems, saw this new innovation of Tesla a financial threat and refused any further loans. Also Morgan had Tesla "blackballed" from receiving loans from other bankers.

Eventually, the Wardenclyffe project was repossessed because of Tesla's outstanding debts. Discouraged, Tesla went on to invent smaller more efficient systems. Tesla invented a vertical takeoff air craft which was patented and later developed an electrical antigravity craft that was not patented. This electrical propulsion system used a combination of high voltage direct current and alternating current fields.

In more recent times, John Hutchison experimented with two Tesla coils and a Van de Graaff generator which created interference zones in the high voltage A.C. and D.C. electric fields. In this interference zone, strange things would start to happen, including levitation of heavy objects, the molecular shattering of steel objects and materials becoming embedded inside of other materials. These experiments, which have been photographed, videotaped and seen by many, lend credibility to Tesla's electrical antigravity craft and the effects of the Navy's ill-fated Philadelphia experiment.

In 1895, Tesla also discovered time alteration effects when he was almost electrocuted from a high voltage experiment. He was experimenting with a highly charged rotating magnetic field when he was accidentally hit by an arc of over 3 million volts. Not only was he paralyzed by the high voltage, he was also found himself outside the normal time reference frame. He could see the past, present and future all at the same time!

Luckily, his assistant shut off the power to the experiment saving Tesla's life. Tesla was intensely interested in this experience after he recovered from the shock and conducted a series of more experiments with an, eye to greater safety, that led to his discoveries of teleportation and time travel.

About this time Tesla was approached by members of the secret NYMZA organization of New York that oversaw the secret Aero Club out west in California. Tesla may have shared the secrets of his electrical antigravity craft with them. If so, it all was kept pretty secret.

Many of his discoveries were not patented because Tesla correctly thought that they were too dangerous for most of spiritually un-evolved mankind.

He did think however, that the United States and England were more civilized that most of the world and offered their governments his anti-war machine which could create a protective shield around their countries that enemy bombers or artillery could not penetrate. Thomas Bearden has written on this "Tesla Shield" which is created with longitudinal wave interference zones in his book, *Fer De Lance* and other of his writings. Also, Tesla offered his Death Ray to the U.S. Military who tested it, not realizing how powerful it was, and blew off the top of a mountain in New Mexico with it.

Tesla also worked on the "Philadelphia Experiment" (Actually called Project Rainbow) for the U.S. Navy during World War II. The navy wanted to make ships become invisible so that German U-boats wouldn't be able to torpedo them. But, experiments in invisibility began before World War II.

The research for invisibility was started at the University of Chicago around 1931, under the guidance of the University's Dean, Dr. John Hutchinson Sr., Nicola Tesla and Dr. Kurtenaur. The project was moved to Princeton's Department of Advanced Studies in 1934 and Einstein and Dr. John Von Neumann joined the group. In 1936, they were able to make small objects disappear. This caused the Navy to become encouraged and the Navy supplied more funding for the project.

In 1937, Tesla was accidentally hit by a cab and was forced to resign from the project and John Von Neumann became head of the project. By 1940, the Navy was ready for a full test on a small ship at the Brooklyn Navy Yard. Power was supplied from nearby ships and everything was operated by remote control with no crew on board the test ship. The small ship disappeared and the project was declared a success. After that, the Navy's funding of the project increased dramatically.

However, the next test was on a battleship, with a crew on board that would require tremendous magnetic fields to cause disappearance. Tesla, who was now recovered from his accident and acting as a consultant, worried that the magnetic field strength being used to cause the ships disappearance would have harmful effects on the ship's crew members and wanted more time to develop a safer system. Von Neumann did not agree. The Navy wanted to push ahead because it was a war time emergency. So, Tesla had the ships coils de-tuned so that the experiment would fail and resigned from the project after it did.

The project was then placed solely under Professor Von Neumann, who later used the smaller, destroyer escort, Eldrige to do the experiment on. On August 12, 1943, the experiment succeeded in making the Eldrige disappear with disastrous results for the crew as can be read about in the many books on the Philadelphia Experiment.

It seems the Philadelphia Naval yard has also been used for other interesting experiments. A friend of mine, Edward Callahan, related to me the following story:

> "When I was about 4 years of age, my father, Edward Callahan, a World War II veteran and POW that was awarded a Purple Heart, took me to an Open House at the Philadelphia Naval Yard on Armed Forces Day in 1953 or 1954. I remember an aircraft carrier and submarines in port. We met my Uncle, Frank Cassidy, my father's cousin, dressed in uniform to have lunch.
>
> Before, lunch at the officer's club, we were driven to the Naval Air Base on the opposite side of the base to a hanger. The driver took us to a side door of the hanger to a viewing room. A Marine Master Sargent, in dress

wools, led us to a viewing room. I remember that he had 6 hash bars on his uniform, as my father was a Sargent when shot and captured.

Through the windows of a viewing room, we saw a disk shaped object, about 30 feet in diameter, anchored to the floor with large chains, floating about 10 to 12 feet off the ground. Men in fur parkas with lab coats under were all over the craft with clip boards using ladders to climb up on the ship.

My uncle Frank told me that 3 passengers were aboard when the craft crashed and was captured. Two died in the crash and one was still alive. I was told that they were about my size, 42 to 48 inches tall. I felt bad about the ones that died and worried about the one that survived.

I remember the craft having a bluish light on the bottom side releasing a steam like vapor. The craft was a grey/blue color, shimmering and changing but dull in shade.

We left after about half an hour and had lunch. After lunch, my uncle left us. My father and I went back to the Open House and saw a room full of computers the size of refrigerators. I got to see myself on television to my delight. My father was not at ease and we returned home. At home, I told my mother all that I saw. But, my father denied my story"

Young Edward was dismayed that his father would lie like that. He didn't realize that he and his father had been allowed to see a highly classified project that would have to be kept totally secret. Ed Callahan also informed me that both his father and Frank Cassidy were Jesuits and that Frank Cassidy was a

Colonial in the New Jersey National Guard. That could explain how they got a security clearance to see the secret project.

The men working on the recovered disk were wearing fur lined parkas because it was cold in the room where it was. As we shall see later, some forms of antigravity and free energy technology do take energy out of the surrounding medium causing it to cool down.

In any case, the many discoveries by Tesla were pretty amazing for the people of that time. The German government placed spies close to Tesla to discover some of his secret technology. George Scherff was one of these spies who became Tesla's trusted assistant for many years.

George Scherff was actually George H. Scherf from Dölitzsch, a small village south of Leipzig, Germany. George Scherff altered his name on immigrating to the U.S.

Scherff also had another alias - Prescott Sheldon Bush. His son, George Scherff Jr., also known as George Herbert Walker Bush, would later become the Director of the CIA and President of the United States. (3)

Scherff/Bush was also an "auditor" with the Union Sulphur Company. The company's "president," German chemist Herman Frasch provided Scherff/Bush with the perfect opportunity to advise him about Tesla's patents and their financial worth, as well as being a conduit to the oil-rich Rockefellers, for whom they both worked.

Using his double identity, Scherff/Bush stole Tesla's inventions, sold them to the Union Banking Corporation (UBC) through his vice-presidency and directorship under the alias of Prescott

Sheldon Bush, to be forwarded to Paul Warburg (banker), Fritz Thyssen (industrialist) and I.G. Farben (the largest conglomerate of chemical companies in the world) — the latter two being industrialists loyal to Hitler's Nazi Party. Some of Tesla's discoveries would later be used by the Third Reich in their secretive flying saucer program.

Tesla's electric air ship was based on his Dynamic Theory of Gravity which utilized Aether. According to Tesla, Aether fills all space. Matter is the result of vortices in this Aether, spinning at the speed of light. Aether is comprised of "carriers immersed in an insulating fluid". Momentum can be transferred by the electromagnetic field according to Poynting's Theorem. All of Tesla's theories were based on his actual experiments. If a static high voltage field is combined with a high voltage high frequency alternating field, interesting effects begin to happen. These effects were used to propel Tesla's electric air ship. These effects are also discussed in *Occult Ether Physics* by William Lyne.

The Nazis weren't just using Tesla's and Thomas Townsend Brown's discoveries in their saucer project. Another scientist who had developed antigravity technology was Victor Schauberger. He studied vortexes in nature and discovered many amazing things about vortexes and developed free energy and antigravity devices from his vortex implosion technology.

When a moving fluid is forced into a pipe of lesser cross sectional area, it has to speed up. When a fluid is rotated in a vortex inside a pipe, it experiences less friction as it flows through the pipe than a non-rotating fluid. When a rotating fluid is forced to rotate at a smaller radius, conservation of angular momentum causes the fluid to speed up. The center of a fluid vortex is cooler and the periphery is hotter. Electrical charge separation takes place in fluid vortexes.

Shauberger used these fluid principles to design his free energy devices, using spiral pipes with decreasing cross sectional area to rotate the fluids and cause them to increase in velocity, in spiral pinwheel like devices that terminated with tangential nozzles that would cause the rotation of the device. Then, the exiting water was pumped back to the top entrance to repeat the process. Centrifugal force would force the fluids through the device as it rotated.

An electric starter motor would rotate the device until it reached its operational R.P.M. At that rotational velocity, the device would become self-powering. Higher RPM would cause a runaway situation where the device would keep speeding up until it destroyed itself. So a speed regulating device was also incorporated. Above the operational RPM, the device would also power an electric generator attached to the rotational shaft. This was the physical principle of Schauberger's free energy generator.

While he was experimenting with devices using air in place of water, he actually discovered a way to develop levitational forces, which he incorporated into another invention which he called a "Repulsine." (4)

In May, 1941, Shauberger and his patent attorney were approached by the Nazi SS and told that he could continue his work but from then on it had to be done in secret. Later, Shauberger would write his son Walter that he was working in Gablonz, Czechoslovakia (present day Jablonz on the Czech, Polish and German border) and what he was doing was secret.

Most of Germany's flying saucers were still under development by the war's end and not regularly used in combat. Many were destroyed so as not to fall into enemy hands and others were flown away to secret bases in Antarctica and Argentina. However, some of the saucer technology and the technicians that developed them were captured by the U.S. and Russia and has been kept secret from the public ever since.

In the 1920s, Thomas Townsend Brown experimentally discovered a connection between gravity and high voltage electric fields. A theoretical explanation of this effect required Aether for it to work. If a high dielectric constant material was placed between two plates, forming a capacitor, and a high voltage was applied to these plates, the whole capacitor would experience a force in the direction of the positive plate. The voltage required was on the order of 30,000 volts or higher. The whole capacitor was encased in anti-corona material to prevent leakage current. He called these units "Gravitators."

In 1922 Brown, at the time a student at the California Institute of technology, had a hard time convincing his science professors of his discovery. Disappointed, he ended up moving closer to his home in Zanesville, Ohio to another university. At Dennison University in Granville Ohio, Brown befriended Professor Dr. Paul A. Biefield who was quite interested in brown's electro gravitational experiments. They experimented together on this effect, which later became known as the Biefield-Brown Effect.

Many mistakenly believe that the Biefield-Brown Effect force is due to an ion wind. However, experiments demonstrate the effect works even better in a vacuum, where there is no ion wind. The effect, using solid dielectrics increases with more massive dielectrics and with higher dielectric constants of the mass. The force seems to increase as the square of the voltage.

There was no measurable reaction force to the Biefield-Brown force. So, it violated Newtonian principles. Some theorists posit that the reaction force is on space itself - essentially pushing against space. Brown thought that an artificial gravity field was produced which caused his gravitators to "fall" or be accelerated in the direction of the force.

However, when the dielectric becomes fully polarized, which usually takes several seconds (depending on its dielectric relaxation time), the effect ceases. Then, the voltage needs to be turned off for a while (usually several minutes) for the dielectric to become un-polarized. Then the voltage can be turned back on and the process repeats. This was because the polarized dielectric caused an equal and opposite electric field to the one caused by the electrified capacitor plates.

Brown developed multiple gravitator units that could be switched on and off sequentially to make continuous force possible. These units were placed in model trains and boats and tested and proved quite practical as a propulsion technology. On the boat, it was noticed that the water closest to the hull was also moved forward by the Gravitators, considerably reducing the boat's friction traveling through the water.

In another embodiment, Brown placed Gravitators around a wheel which would turn continuously when the high voltage was applied. After attaching an electrical generator to the shaft of the wheel, he found that he could generate far more electric power that was being consumed by his high voltage generator. In 1927, Brown applied for a British Patent which was issued to him in 1928 (British Patent Number 300,311).

Townsend Brown developed a means of propulsion that bypassed the dielectric polarization problem by using air as the dielectric and allowing air to flow through his saucer like device. Thereby, polarized air molecules were replaced with un-polarized air molecules between the high voltage electrodes and the force would be continuous.

Brown also discovered that the force was enhanced by using unsymmetrical electrodes and got the best effect when using a thin wire as the positive electrode and a flat plate or foil as the negative electrode.

This concept was patented by Brown with U.S. patent Number 2,949,550 issued on August 16, 1960 titled Electrokinetic Apparatus. Now days, hobbyists around the world are building "Lifters" based on this idea. Again, many falsely believed that Lifters rose off the ground due to ionic wind. And again, it was proven that Lifters worked even better in a vacuum with no ionic wind.

Brown also set up a lab at his home in Zanesville, Ohio and did research at the Swazey Observatory. He noticed that the gravitator forces were also affected by planetary positions, with the sun and moon having the greatest effects, the most pronounced effect during solar and lunar eclipses. Also the Earth's position relative to the galactic center had an effect. These experiments further demonstrated that the Biefield-Brown effect is related to gravity, namely the tidal effects of other bodies in space.

Another electrokinetic effect was discovered by other scientists, Tesla noticed that high voltage direct current, when pulsed through rotating spark gaps would throw out a certain type of energy which Tesla called radiant energy. It was a different phenomenon than alternating current. He could reduce the pulse width by using magnetically quenched spark gaps. Sufficiently high voltage and short pulse width would create a stinging sensation on Tesla's skin caused by this radiant energy, which seemed to be carrying some type of momentum. The radiant energy was difficult to shield and seemed to pass through most materials including conductors. However, the conductors would quickly become highly charged with electricity.

Varying the frequency of the pulses would vary the effects. Around 10,000 pulses per second the stinging sensation was no longer felt but, a warm sensation would fill his body. At even higher pulse rates nothing was felt at all. Tesla developed the high voltage pulsating D.C. technology into highly efficient energy multiplying and transmitting devices.

At Colorado Springs, he developed a flying platform powered with pulsating D.C. which derived its power from his powerful magnifying transmitter. Occasional ranchers in the area would see Tesla flying around in his electric aerial vehicle.

Oleg D. Jefimenko has done much theoretical and experimental work with high voltage static electric fields, electromagnetism and gravitational fields. In his book *Causality Electromagnetic Induction and Gravitation,* he shows shortcomings in present electromagnetic and gravitational theory and presents the Electrokinetic electric field which equals the negative of the time derivative of the Magnetic Vector Potential. This field is parallel to a wire carrying the current and the strength depends on the quickness of the current pulse. The faster the rate of change of the current, the stronger would be the Electrokinetic Field.

This finding is reinforced by experiments done by German scientist, Rudolf G. Zinsser where pulsed electricity is conducted to two parallel plates immersed in water that would generate a force on the whole assembly in the same direction predicted by Jefimenko. The water had a dielectric constant of about 80, the pulses were a few nanoseconds in duration, but the force was sustained and would last even for a while after the pulse generator was turned off. This became known as the Zinsser Effect.

What these scientists had discovered was that pulsing high voltage D.C. current would entrain the Aether of Space and cause it to flow, allowing not only electrical effects but also mechanical effects under the proper conditions.

The reason for calling this chapter "Secret Science" is quite simple. I majored in Electrical Engineering with a minor in Physics at University of California at Santa Barbra and received my BSEE degree in 1982. At no time was any of the above subject matter taught at the University or even mentioned.

In one Electrical Engineering class, the professor stated that longitudinal electromagnetic waves were mathematically proven to be impossible. Having privately studied Tesla's inventions, I quickly raised my hand at this statement. The professor pointed to me. I said that Nicola Tesla used spherical antennas that could only produce longitudinal waves. The professor waved me away with disgust, stating that he didn't want to talk about Tesla. This dampened my ideal of Universities being places for the free exchange of ideas.

But, besides Universities pushing well-ordered, undeviating programs that corporations want them to push, there were other factors involved in having a secret science. The technology of free energy and antigravitation would give the masses too much freedom.

Imagine having your own personal flying saucer that was powered with free energy and that you could fly anywhere in the world you wanted. Imagine having no electric or heating bill because your house is powered by free energy. It would be impossible to have border check stations or even borders. Ports and airports would become obsolete. The Oil companies would have to limit their activities to chemical material productions like plastic lumber instead of wastefully burning oil for fuel. Imagine houses built with plastic lumber that would last hundreds of years instead of decades due to rot and termites. Look at the rainforests this would save!

The elite that control the world governments from behind the scenes would lose their cash flow and control. Also, there was the military factor. Much of this secret science has been weaponized and is classified under the National Security umbrella.

One, underlying principle of this secret science is the concept of Aether. It is falsely taught that the Michaelson-Morely experiment disproved the existence of Aether. Lorentz developed his contraction formula to explain the negative results of the Michelson-Morely experiment while allowing the Aether to exist. Einstein took Lorentz's contraction formula and used it in his Special Theory of Relativity which still didn't disprove the existence of Aether. Later, Einstein developed the General Theory of Relativity which explained gravity as a 4 dimensional curvature in space-time. Nowhere in this development was Aether disproven.

After Einstein developed his General Theory of Relativity, he realized that Aether was necessary. In his 1921 monograph *Sidelights on Relativity* Einstein states: "Recapitulating, we may say that according to the general theory of relativity, space is endowed with physical qualities; in this sense, therefore, there exists Aether. According to the general theory of relativity space without Aether is unthinkable; for in such a space there not only would be no propagation of light, but also no possibility of existence for standards of space and time. (measured rods and clocks), nor therefore any space intervals in the physical sense."

Furthermore, the Michelson-Morely experiment was conducted underground where the Aether may well have been stationary relative to interferometer, in which case no Aether wind could have been detected. Other Aether wind measuring experiments, like those conducted by George Sagnac, Ernest Silvertooth and Dayton Miller have measured a variable speed of light, indicating an Aether wind. The anti-Aether theorists conveniently seem to forget about these other experiments.

Another point to consider is that all of Maxwell's equations were developed using Aether dynamics. To throw away Aether, to be logically consistent, one would have to also throw away Maxwell's theories and equations.

At one time, people thought that air didn't exist. To them, wind was a mystery. In like manner the "empty space" people think that Aether doesn't exist. So called "empty space" has properties - a constant of magnetic permeability, a constant of electric permittivity, and an impedance. These properties of "empty space" determine, among other things, the speed of light. A no-thing has no properties. A something has properties. The Aether in "empty space" is the something that has the above mentioned properties.

Those who wanted to keep anti-gravity and free energy science secret realized the best way to do so was to eliminate the concept of Aether. That is why we are taught the illogical concept of empty space, that has no mechanization to transmit light, magnetic force, electric force or gravity across the vast reaches of inter stellar space or action at a distance, in our Universities.

Dan A. Davidson, in his book *Shape Power*, has shown many

interesting experiments that clearly demonstrate the existence of Aether and yields explanations for subtle phenomena like pyramid power for which conventional science has no explanation.

One of the more dramatic experiments described in the book involved the Joe Parr Gravity Wheel. This wheel consisted of 24 two dimensional pyramid shaped copper pieces placed on an insulating wheel which is spun between low gauss (100 gauss) magnets. This set up would generate an "energy" bubble around the whole apparatus. This bubble could be enhanced by injecting negative ions into the bubble.

Inside these energy bubbles, energy sources like radioactive isotopes, electromagnetic transmitters and so on, would have the intensity of their radiation reduced. Gravity was also reduced! Sometimes the weight reduction was more than the apparatus weighed.

It was also discovered that the weight loss correlated with celestial alignments and that there seems to be energy "conduits" between the Sun and certain stars. For example there was discovered an energy conduit between the Sun and Orion. When the Earth passed between the Sun and Orion, it passed through a conduit which caused the Joe Parr Gravity Wheel to try to move down the conduit!

Davidson and Parr are refining these experiments with the goal to develop a space drive.

Thomas Bearden, Ph.D. points out the many flaws in modern Electrodynamics. It is worth repeating his 2002 Overview and Background statement in the 2002 edition of his book *Fer De Lance*:

"For some time Russia and several other nations have possessed highly advanced 'extended electromagnetics' (energetics) weapons of a very novel kind, using a dramatically extended electrodynamics theory. To comprehend these weapons requires a combination of non-Abelian electrodynamics in at least the O(3) gauge symmetry, general relativity, Bhom's interpretation of quantum mechanics and the use of time domain (scalar) energy, fields, and potentials to directly enter distant 3-spatial points without propagating energy through 3-space. It also requires correcting many of the serious errors in classical electrodynamics. Probably the most elegant and applicable extant model to deal with these phenomena and weapons is Sachs' unification of general relativity and extended electrodynamics, particularly as implemented in O(3) electrodynamics. O(3) allows direct engineering by EM means…"

For those of you that haven't been schooled in group theory and advanced physics, Bearden is, among other things, talking about using higher dimensional theory to transport energy (and therefore theoretically possibly matter) - not through space - but through the time domain to have it instantly appear somewhere else.

In fact, Bearden later explains the solution to the source charge problem of charged particles by using the time domain.

Supposing that electric charge was a property of Aether (a virtual particle Aether according to Bearden) flowing into or out of a charged particle. The source charge problem is where does the Aether come from, in the outflowing case and where does it go in the inflowing case? Bearden's answer to this question is that it comes through the time domain to the negative 3 charge, thence to the positive 3 charge, thence back to the time domain in a 4 space circulation.

He further states that no observable actually exists in time. The observable only occurs in 3-space. This is like a movie film which is composed of still frames or "observables" which we actually observe one frame at a time. The motion of the still frames through the projector creates the illusion of motion. The observation only occurs through the process of taking a time derivative of a 4-space event. This leads to the mistake of thinking a 3-space observable as a 4- space cause. A separate 3-force does not act on a separate 3-space object. No such situation exists in nature, and nothing moves or changes in 3-space alone. For an object to exist in time it requires at least a 4-space because a time length is required.

In 1921, Theodor Kaluza developed a theory that unified electromagnetism and gravity that took 5 dimensions. Gravity and electromagnetism are caused by a 5 space G-field in this model. The 5 dimensional component of this field is electromagnetism and the 4 dimensional component is gravity. As an example, between two electrons, the 5-space G-potential bleeds off into an electrical force field ten to the 42 power times as powerful as the 4- space G- field.

At first, Kaluza's theory was virtually dismissed because it predicted many unknown potentials and fields. But by the 1970's it was reinvestigated when it was discovered that an 11 dimensional version of his theory predicted many experimental outcomes.

Einstein's curved space-time according to Bearden means virtual particle (Aether) pressure or flux density changes from one place to another. This virtual particle pressure is a scalar since it has no vector direction.

There are a number of scientists that think that gravity is a "push" rather than a "pull". They posit that there is a cosmic flux or wind that exists throughout the universe. Some claim that it is a neutrino flux. This flux is random coming from every direction.

So, far in space there would equal pressure (although much less than atmospheric) on a body in all directions with no resultant overall force. However near a massive body such as a planet, some of the flux is absorbed in passing through the planet, creating an unbalance of the flux and a lower pressure on the side of the body facing the planet.

The unbalanced particle flux now creates an unbalance pressure on the body, forcing it towards the center of the planet. This is the force we call gravity. This concept is discussed in Dr. Hans A. Nieper's book *Revolution in Technology, Medicine and Society: Conversion of Gravity Field Energy*.

Some scientists believe that "Dark matter is Aether. Gravitational force could as easily be caused by an Aether flux acting in a similar manner if there is some interaction between Aether and matter. And, many scientists believe that, however small, such an interaction does exist.

Thomas Bearden, in his book *Fer De Lance*, states that the Russians have weaponized "extended electromagnetics" and explains just how it is done. One modifies a radar transmitter to transmit longitudinal EM waves instead of transverse EM waves. A second radar antenna is also modified the same way. The two antennas are spaced apart and their radar beams are directed to a distant point. At the intersection zone of these two radar beams, interference between the two longitudinal beams will set up a stationary zone of normal transverse waves or a localized bottle of EM energy. This process is called Longitudinal Wave Interference (LWI).

Energy can be pumped into or out of this interference zone depending on the biasing with reference to ground of the transmitting antennas. In the straight forward version of this weapon, one can pump heat into or out of the interference zone using LWI, causing intense heat or cold in the target volume. Rapid pulsing of the biasing could cause heat or cold explosions in the target volume.

In more complex versions certain "zero summed vectors" can be modulated into the radar signals to produce different results, like weakening of metals, increasing or decreasing the mass and inertia of matter, transmitting diseases, or remote mind control.

If you applied a number of forces to an object that vector summed to zero, the object would have zero net force and would not accelerate. In normal physics, the vectors summed to zero and therefore could be dismissed. In extended physics there was still a pressure or a tension on the object that could not be simply dismissed.

Likewise in extended electromagnetics, if the E vectors or the B vectors summed to zero, there is still a stress on space time (or Aether or virtual particle flux). This stress is the gravitational potential which can be engineered to be positive or negative, depending on the biasing of the transmitting antennas, (gravity or antigravity). Using LWI, and zero summed E and B vectors of sufficient intensity, distant objects could have their mass and inertia changed. Objects with zero inertia can undergo extremely high accelerations with little force, as many have observed UFOs to do.

Scalar Electromagnetic waves are different than Transverse EM waves which, if we use the "z" axis as the direction of wave travel, the E and B fields oscillate about the "x" and "y" axis and the longitudinal EM wave that has the E vector oscillating about the "z" axis, in that the Scalar EM wave oscillates about the time axis ("t") with no vector direction in space, hence the name "scalar." Since the Scalar EM wave doesn't exist in space, it can instantly transmit energy to a distant location via the time dimension.

A Radar can easily be converted to transmit Scalar EM waves by transmitting 2 equal EM signals that are 180 degrees out of phase. The EM vectors will cancel each other but an Aether pressure wave or scalar wave will result. This wave will also be a tempic or time wave and a gravitational wave.

If two or more Scalar EM radar beams are intersected an interference zone is created, as with the longitudinal EM wave case, transverse EM waves will appear in the interference zone. The advantage of using scalar EM, besides the ease of converting normal transverse EM radar to scalar EM radar, is the instant appearance of this trapped EM energy in the interference zone with zero speed of light time delay.

Tremendous amounts of energy can be extracted from the Earth's molten core using this technology with the proper biasing of the antenna system. According to Bearden, the Russians have developed systems that transmit directly through the Earth to distant targets with their large scale antenna systems. Tesla also proved that he could send longitudinal electromagnetic waves through the Earth. Bearden also claims that the Russians can change the weather and trigger earthquakes with this technology. One can't help but wonder if the U.S. HAARP system can also do these things.

The High Altitude Auroral Research Project, started by DARPA was at first operated by the Office of Naval Research then turned over to the Air Force. It consists of a thirty acre antenna farm at Gakona, Alaska. The antenna farm acts as a huge phased radar antenna system that is steerable and can pump microwave energy into the ionosphere. This energy can heat the ionosphere and alter the shape of the ionosphere. This much is admitted to by the Air Force. Also the microwave signals can be deflected off the ionosphere and steered to remote locations and used as an over the horizon radar.

By slight modifications the EM radiation from HAARP can be altered to be scalar waves or longitudinal waves and used as similar weapons that Bearden says the Russians have.

The Air Force plans to dismantle the HAARP facility in 2014, claiming that they got good service from the project to control the ionosphere and that it was only a temporary project. Many researchers think that the technology has been reduced in size and improved into the floating platform SBX-1 X-Band, phased array radar system. These systems can be towed to any location on the planet's oceans.

According to Bearden, extended electromagnetics has many potential uses including electronic communication between under sea submarines, instantaneous communication in deep space projects, free energy technology and it could be engineered for hyper spatial travel, including teleportation and time travel.

An interesting experiment carried out by an associate of Thomas Bearden, named Floyd Sweet involved a specially conditioned magnet. This magnet was conditioned so that its magnetic domains would oscillate at 60 cycles per second when activated by a coil with a 60 cycle input applied. Another coil at right angles to the activator coil would, by induction from the oscillating magnet, generate electrical power, which was much greater than the power input by the activator coil.

This free energy device seemed quite simple. The proprietary secret was in the process to condition the magnet so that its domains would oscillate. Unfortunately, this secret, Sweet would take to his grave.

In any case, Sweet's generator had the interesting property that as you added more load to the output it would cool down. Conventional generators would have heated up as load was added. This property caused Thomas Bearden to think that negative energy was involved. And, if negative energy was involved, the weight of the device theoretically should also decrease as load was added to the device.

So an experiment, described in Thomas Bearden's book *Energy from the Vacuum*, was conducted to test this theory out. The device was placed on a scale and load to the output was incrementally increased. A plot of weight verses load was recorded. And, as Bearden thought, the weight decreased with increasing load!

The experiment was done over the phone with Bearden at one end and Sweet at the other. Previously, Sweet had his devices become unstable and actually explode when loaded too highly. So, Bearded warned him not to take the weight down to zero. But, the experiment brought the weight down to almost zero. If it wasn't for the mechanical instability problem with the magnetic material, they could have probably taken the weight into the negative region and it probably would have levitated.

2

The Science and Technology of Free Energy

Free energy is quite a simple thing. Free energy machines have existed for centuries. A free energy machine simply extracts energy from the surrounding medium and puts it to some practical use. Typical free energy devices are sail boats, water wheels, windmills and so on. The idea is to tap the energy of the medium of moving air or water in these cases. But, there are other mediums from which energy can also be extracted.

In the 20th century, free energy machines became more sophisticated. Here, I intend to discuss the little known more sophisticated free energy machines.

In our atmosphere at Standard Temperature and Pressure (STP), the molecules of air have an average velocity of the speed of sound. The velocity of these molecules is totally random with zero net velocity. So, there is no 767 MPH wind. These random air molecule velocities do however possess considerable kinetic energy which manifest as air temperature and pressure. There was a businessman who developed a way of extracting this energy content of air to produce electricity.

Dennis Lee had a heat pump factory in the State of Washington.

When President Carter passed an energy efficiency tax credit law, Dennis Lee asked his engineers to develop a more efficient heat pump so his company could qualify for the tax credit. The typical heat pumps of the time had a Coefficient of Performance (COP) of about 4.

A COP of 4 meant that the heat pump could pump 4 times as much heat energy as the electrical energy it consumed to operate the pump. Dennis Lee's engineers eventually developed a heat pump that was tested in independent laboratories with a COP of over 12! With that high of a COP, Dennis Lee asked his engineers if it would be possible to take the heat pump output and run a heat engine that could power a generator and get more electricity out than was put in.

The engineers stated that the problem with that idea was the Carnot efficiency of heat engines which typically was below 30%. One engineer however knew of a new type of heat engine developed by a Dr. Fischer that was theoretically 90% efficient. This engineer also knew how to contact Dr. Fischer.

Meanwhile, the Washington State electric utility company did not like the fact that Dennis Lee's company was now selling such efficient heat pumps because they were losing money on all the homes that were using the more efficient heat pumps. They used their influence with the Washington State Attorney General's office to bring a contrived lawsuit against Dennis Lee's company.

Dennis Lee simply moved his operation to Ventura County in California. By now, the company had teamed up with Dr. Fischer and developed a Fischer cycle heat engine. Soon, they had perfected their machine to make electricity from the air. They rented an auditorium in Oxnard, California to demonstrate their device. About 200 attended the demonstration including mechanical and electrical engineers and members of the press.

On stage, the machine was displayed, complete with heat pump, Fischer cycle heat engine, generator, radiators and a panel with 100 light bulbs of 100 watts each. Skeptics from the audience were invited to go up on stage and examine the set up to show that there were no hidden wires and so on.

The machine was started with an extension cord from the theater's electric power. Once the machine got up to operating RPMs the power was switched to the generator and the extension cord was disconnected and removed from the area. Now, the machine was powering its self.

Then, Dennis Lee threw another switch and the panel of 100 100 watt bulbs became brightly lit. That represented a load of 10 kilowatts. Denis stated that this was not free energy because his machine was simply extracting the heat energy in the air to power the load and that conduction currents would soon cause the air in the auditorium to cool down. So, not only was his machine producing electricity but it was also an air conditioner!

The only reason we don't have this technology on the market is because of suppression by the energy companies and their influence over our government.

Two days after Dennis Lee's successful demonstration in Oxnard, his laboratory and business office was raided by the Ventura County Sheriffs. A semi-truck and trailer unit was backed up to his Laboratory and all the equipment inside was confiscated and loaded into the trailer. All the research papers and business contracts were likewise removed from his office and Dennis Lee was arrested.

He was charged with failing to file some obscure document with the Ventura County Clerk's office. Normally, this kind of a problem would simply be solved by the County Clerk mailing a letter requesting said document. So, this was obviously, a bogus arrest. Dennis Lee was held for a year without trial. Then, after his company had folded, he was released and all charges dropped. Dennis Lee was never able to get his confiscated property back because the Sheriff's office claimed that they were holding it for evidence!

Another experimenter, after hearing about the Dennis Lee concept, made a solar heating panel about 10 feet by 10 feet in area and connected the hot water output to a heat pump to step up the temperature further and was able to run a 350 HP steam engine from the small solar panel and heat pump combination.

If we accept that there is a finer medium than air surrounding us, which is known as Aether, and in which the particles have an average velocity of the speed of light, we might see even better possibilities for extracting energy from this medium.

The problem is in discovering something that interacts with Aether. While air can be compressed, Aether would pass right through every conceivable material. So, an Aether compressor or engine would be impossible to manufacture.

The solution is in electromagnetic phenomena. Electric and magnetic fields do interact with Aether. In fact, these fields might just be different kinds of Aether flows. The electric field might just be a linear Aether flow and the magnetic field, a vortex Aether flow circulating around a charged current flow. It is known that electric and magnetic fields, that are time variable, change into each other.

To those that still believe that in a vacuum, space is totally empty and there is no Aether, I would point out a property of space known to physicists as the "polarized vacuum".

If two conducting plates are separated and placed in a perfect vacuum and a high voltage is placed across these plates, the vacuum becomes polarized. The higher the potential difference between the plates, the greater is the polarization of the vacuum between the plates. When the voltage is raised to a sufficiently high level, electron positron pair production starts occurring in the vacuum between the plates!

This well-known phenomenon demonstrates that this vacuum has something in it and is not empty space. Polarization only occurs in substances, however infinitesimal, with electric charges that are separated in space. And, the electron positron pairs didn't just evolve out of nothing.

Many experimenters were fascinated with permanent magnets because they continually out put a magnetic field without any input of energy. After much experimentation, a number of motors running off of permanent magnets were developed, the most famous being the Howard Johnson magnet motor, US patent # 4,151,431.

The 1980 spring issue of *Science and Mechanics* magazine ran a story on Howard Johnson's magnetic motor and information can be seen here: http://www.newebmasters.com/freeenergy/sm-pg45.html

Howard Johnson was not the only one to invent magnetic motors. Other inventors like Troy Reed and Perendev, and companies like Cycclone Magnetic Engines Inc. and Terawatt Research LLC., to mention a few, also have come up with various versions of magnetic motors. So, the concept is quite real. But, by various means, these inventions have been prevented from entering the market place.

Other methods of tapping the power of magnets also exist. One simple way is to pulse an electromagnet coil wrapped around or close to a permanent magnet. When the electromagnetic pulse reinforces the magnetic field of the permanent magnet field, the coil output power is greater than the input power. Bifilar coils are also useful in this experiment one being the input and the other being the output coil.

One free energy machine using this method is actually on the market in Canada! How they got past the suppressors, I don't know. One version of the machine has 4 outputs, each putting out 100 amperes at 12 volts or a total of 4.8 Kilowatts. Also, there are more expensive versions of greater power. One needs other equipment to be used in conjunction with these machines, like storage batteries, inverters and charge controllers just like with solar systems.

The inventor is Richard Willis and the company is Magnacoaster. Some customers have complained about the long delivery time after placing orders. This is probably because these machines are hand built - not mass produced. But if you are curious, their web site is: http://www.magnacoaster.com

Interesting things happen when two electromagnets are placed so that the same poles face each other and the coils are pulsed. Vectorily, the two magnetic fields cancel. However, there is a stress placed in the space between the poles. This stress creates a pressure in the Aether and creates what is known as a scalar electromagnetic field, because pressure is a scalar quantity. Also of course, a repulsion force between the electromagnets takes place.

Edwin Grey developed a pulsed capacitive discharge motor which used this concept. High energy storage capacitors were charged up to 20,000 volts by rectifying the output from auto ignition coils. This 20,000 volt charge was then pulsed through rotary spark gaps through rotor and stator coils in his motor. The rotary spark gaps were synchronized with the rotor and stator coils so that the discharges took place when the coils were opposite each other, resulting in a tremendous repulsion force on the rotor. This motor was patented as US patent number 3,890,548.

A hundred horsepower Edwin Grey motor was installed in a test car. This car had a range of 700 miles from the charge in just two 12 volt car batteries! This test made the newspapers. Edwin Grey received a certificate of merit from then governor of California, Ronald Reagan.

Another aspect of Edwin Grey's motor was that he was using a form of electricity that was similar to radiant energy discovered by Tesla which Grey called "Cold Electricity". The Cold Electricity device was patented as US patent number 4,595,975. These aspects of Grey's cold electricity are discussed in Peter Lindemann's book *The Free Energy Secrets of Cold Electricity*. Peter Lindemann has a free energy website here: http://www.free-energy.ws

The Evgrey Motor Company was created and people were buying up stock. A site for a factory to mass produce these cars was being determined. Then suddenly, a lawsuit from the Securities and Exchange Commission was leveled against the Evgrey Motor Company. This lawsuit effectively shut down the company.

Let's see here now. The Rockefellers have both considerable ownership of oil companies and influence on the Securities and

Exchange Commission. You don't suppose that they may have been worried that cars manufactured using Edwin Grey's concepts may eventually diminish their cash flow do you?

Edwin Grey's concept of pulsing opposed electromagnets was followed up by experimenter Doug Konzen. He was getting interesting results at much lower voltages, typically 12 volts from a car battery which he was open sourcing on the internet. I corresponded with him in the early days of his experiments via e-mail, trying to convince him to use higher voltages like Edwin Grey had used. But, he kept using low voltages. The thing is that, even using low voltage, he was getting over unity (more power out than in) with a closed loop motor generator combination. Doug Konzen's website is here: http://sites.google.com/alternativeworldenergy/

Another type of motor Konzen experimented with is the Rotoverter. The Rotoverter concept was developed by Hector D. Perez Torres who also gave it to the world by open sourcing it. Basically one modifies a three phase, AC electric motor. What is done is to place capacitors in parallel with the motor coils so that they form tuned circuits that resonate at 60 Hz, greatly increasing the efficiency of the motor.

AC electrical generators can also have their efficiency increased in this way. When the motor is connected to a generator that both have been thusly modified, it is possible to get more power out than is put in. See: http://ggtrust.com/energy_rotoverter.html

Another Free energy researcher is John Bedini who developed a free energy machine in the early 1980s. This machine attained 800% efficiency during initial tests. His machine was demonstrated at the 1984 Tesla Centennial Symposium in Colorado Springs, Colorado. Instead of patenting his machine, he also open sourced it. Bedini got his idea for the device from Ronald Brandt, a personal friend of Nicola Tesla.

Bedini also worked in conjunction with Thomas Bearden on some of his many other projects, with Bearden developing the theory and Bedini developing the electronics and hardware. Thomas Bearden himself, patented a free energy machine which he called a Motionless Electrical Generator (MEG) from a design given him by another free energy researcher, Timothy Trapp of World Improvement Through The Spirit Ministries (WITTS).

For more information here are their websites:

John Bedini: http://johnbedini.net

Thomas Bearden: http://cheniere.org

Timothy Trapp: http://www.witts.ws/

The Electric field can also be used to create free energy machines. Thomas Townsend Brown discovered an interesting effect which connected the electric and gravitational field in the 1920s. Whenever a simple parallel plate capacitor is charged to a sufficiently high voltage, a force on the entire capacitor in the direction of the positive plate is created. The voltage needs to be higher than 30,000 volts to be noticeable.

Brown and a professor at his college named Biefield did considerable experimenting with this effect and concluded that it was an electrogravitic phenomena. In other words, the high electric field was creating a "gravitational well" which the capacitor kept falling into. This effect, which is virtually ignored in physics, is known as the Biefield Brown effect.

Later, Brown developed "gravitators", which we will examine in more detail later in this book, made from these high voltage capacitors. Even though these gravitators required a very high voltage, if properly insulated, very small leakage currents were involved. Typically voltages of 300,000 volts and with currents of millionths of an ampere were involved. So, actual power consumed was on the order of a quarter to a half of a watt.

He placed these gravitators in model trains with the positive plates forward as propulsion devices which worked quit well. Gravitators placed in model boats propelled the boats forward as well as the water surrounding the boat, greatly reducing friction.

Then, he placed these gravitators around a wheel so that the forces produced would cause the wheel to turn. Brown found that this wheel could be attached to an electrical generator and generate much more electric power than the gravitators consumed

A very important principle in free energy is that of resonance. In a typical city there may be dozens of radio stations transmitting simultaneously. If your radio received them all at the same time, you would get a lot of unintelligible noise. Nicola Tesla invented the tuned circuit, in which a capacitor-coil combination will only resonate at a single frequency. So, by varying the capacitance with a variable capacitor or variable coil, the resonant condition of the tuning circuit in your radio will change and be able to "tune in" the desired station while rejecting all the others.

A similar situation occurs with tuning forks. If you bring a still tuning fork near a vibrating tuning fork of the same frequency, it will also start vibrating. If you bring a still tuning fork of a different frequency near the vibrating fork nothing will happen to it.

Now the interesting thing is what happens if you bring a hundred still tuning forks of the same frequency near a vibrating tuning fork of that frequency? The correct answer is they *all* will start vibrating. In fact, here we have multiplied the energy of the first vibrating tuning fork a hundred fold!

How much more power will a 100 kilowatt radio transmitter use when 10,000 radio receivers are tuned in, than when only a 10 radio receivers are tuned in to that station? The correct answer is no more power.

Inventor Donald Smith used these resonant principles with Tesla coils to create a free energy device. He fed oscillating electrical energy into the primary winding of a central Tesla coil. Around this central coil, he placed more identical Tesla coils tuned to the same frequency. These coils also started oscillating at the same frequency. He could then take the electrical energy from the primaries of the surrounding coils add them together and multiply the original energy by a factor depending on the number of the surrounding coils. As with most of Don Smith's free energy inventions, the operating principle was quite simple and effective.

A parametric amplifier can be made by either varying the capacitance or inductance of a tuned circuit at the same frequency (or twice the frequency) that the circuit is tuned to. When this condition is satisfied, the power of the oscillations is dramatically increased. This is another phenomenon of resonance.

Some question where the extra power comes from. Some use quantum physics to explain power amplification by parametric amplifiers. But, simply, if one realizes that both the electric and magnetic fields are Aether in motion and this motion is a collective motion in the otherwise chaotic motion of Aether particles traveling at the velocity of light, one could understand where the energy is coming from. The parametric resonance condition coheres a small fraction of the total energy present in in the Aether into these oscillating electric and magnetic fields.

The rate of power amplification is related to the frequency of the parametric oscillator. So, when designing microwave parametric oscillators, care is required that the power doesn't build up faster than the load can dissipate it. These devices have been known to explode from exponential increases in power.

Varactor diodes are typically used in high frequency parametric amplifiers because their internal capacitance changes with applied voltage. Varying the applied voltage across the varactor diode at the resonant frequency of the tuned circuit causes the parametric amplification phenomenon. Usually, a means to shift the frequency of the pulses fed to the varactor diode is used to shift it's frequency from resonance when the power becomes too great. This detunes the parametric amplifier and reduces excessive power buildup.

Another free energy device that uses the principle of resonance and parametric amplification is the Quantum Energy Generator (QEG).

In this design, a rotor made of a good magnetic conductor like steel is rotated between the pole pieces of a toroidal arrangement of four coils. Two opposing coils are in series with a capacitor bank and the other two opposing coils are the power output.

The rotor is connected to a small motor which speed is varied with a Variac. This rotor effectively changes the inductance of the coils by varying the magnetic flux path of the coils. An exciter circuit starts the voltage flowing in the output coils, which through transformer action, induces voltage into the coil capacitor combination.

When the rotor rpms match the resonant frequency of the coil capacitor combination, the circuit is "in tune." The circuit acts as a parametric oscillator and the power is multiplied. And, by induction, the varying magnetic flux induces power into the output coils. The output power is considerably greater than the power to rotate the rotor. Once the rotor has the resonant RPMs, some of the output power can be fed back to the rotor motor to keep the unit self-running. Then, no input power is required to run the QEG. Off grid units would have to be started with a battery and inverter combination.

The complete set of plans for a 10 KW version of the QEG, including parts specification, is being open sourced to prevent its suppression and is available here: http://www.fixtheworldproject.net/quantum-energy-generator.html

Of course, one could also run your car on water, which is a much better fuel than gasoline. Hydrogen has the highest energy density of any fuel and oxygen is the perfect oxidizer. When hydrogen burns, it combines with oxygen to produce harmless water vapor as an exhaust. So, there is no harmful air pollution.

The water molecule is a combination of one oxygen atom combined with two hydrogen atoms. So, all that has to be done is brake the water molecule into its parts, hydrogen and oxygen. One process to do this, is electrolysis of water.

Typically the voltage from your 12 volt car battery is fed to oppositely charged plates immersed in a cell filled with water. Hydrogen gas evolves at one plate and oxygen gas at the other. This gas mixture is then fed into the carburetor. The problem with standard electrolysis is that it doesn't produce hydrogen fast enough. So, the usual hydroxy gas kits sold on the market only work to boost gas mileage rather than to replace gasoline all together.

However some inventors have discovered more efficient ways to electrolyze water into hydrogen and oxygen gas. Pulsing the current to the cells seems to improve the efficiency of gas production. Some inventors claim to pulse the current at the frequency that the water molecule vibrates, creating a resonant situation which literally breaks the water molecule apart much the same way an opera singer can shatter a glass by singing a note that is the resonant frequency of the glass.

Stanley Meyer and Bob Boyce, two well-known experimenters among others, have actually run their vehicles on straight water by using pulsed current electrolyzers. Before building your own, understand that hydrogen gas is highly explosive and you should study the information and advice on this subject that is available in books and the internet by those that have plenty of experience.

One experimenter accidentally replicated Keeley's Aetherealization of water. He was using a quartz column with a 5.08 cm ID and a Barium Titanate ultrasonic transducer at the bottom. Above the transducer an approximately 10 cm height of water filled the quartz tube. An ultrasonic frequency generator and amplifier combination was feeding about 600 watts of power to the transducer. The experimenter started with a signal of 40,000 Hz which caused about three wavelengths in the 10 cm of water. The frequency was gradually increased until, at about 41,300 Hz, the water in the tube disappeared!

The experimenter looked up at the ceiling right above the tube and saw a clean cut hole in the ceiling that also went through the roof! Another experimenter replicated this experiment and found a similar result at a much lower power than 600 watts.

Supposing these experiments were conducted in strong steel containers and the Aetherealized water vapor stored in a strong steel tank. Supposedly a high pressure would be built up in the tank, as Keeley had discovered. This pressure could perhaps be used to operate an air motor to harness the energy of this process. Here is another free energy concept.

There are many more ways to tap the ambient energy of the medium of space, whether one calls it Aether, virtual particles, or dark matter, and convert it into useful energy. I don't have the space in this book to entirely cover this vast subject. If you are interested and want more ideas on this subject check out: http://www.free-energy-info.co.uk/PJKBook.html

3

The Physics of Time

Time itself has long been a mystery. What exactly is time? Modern science merely states that time is a duration and is measured by repeating cycles, like the rotation of our planet, the swing of a pendulum, or the vibration of atomic systems. One rotation of the planet is a day; one orbit of the Earth around the sun is a year and so on.

But other than duration, what other properties does time have? And more importantly, can time be manipulated?

In 1895, while conducting research with his step-up transformer, Nikola Tesla had his first indications that time and space could be influenced by using highly charged, rotating magnetic fields. Part of this revelation came about from Tesla's experimentation with radio frequencies and the transmission of electrical energy through the atmosphere. Tesla's simple discovery would, years later, lead to the infamous Philadelphia Experiment and the Montauk time travel projects. But even before these highly top-secret military programs came about, Tesla made some fascinating discoveries on the nature of time and the real possibilities of time travel.

With these experiments in high-voltage electricity and magnetic fields, Tesla discovered that time and space could be breached, or warped, creating a "doorway" that could lead to other time frames. But with this monumental discovery, Tesla also discovered, through personal experience, the very real dangers inherent with time travel.

Tesla's first brush with time travel came in March 1895. A reporter for the New York Herald wrote on March 13 that he came across the inventor in a small café, looking shaken after being hit by 3.5 million volts, "I am afraid," said Tesla, "that you won't find me a pleasant companion tonight. The fact is I was almost killed today. The spark jumped three feet through the air and struck me here on the right shoulder. If my assistant had not turned off the current instantly it might have been the end of me."

Tesla, on contact with the resonating electromagnetic charge, found himself outside his time-frame reference. He reported that he could see the immediate past – present and future, all at once. But he was paralyzed within the electromagnetic field, unable to help himself. His assistant, by turning off the current, released Tesla before any permanent damage was done.

Tesla, would continue his experiments with space-time manipulation, using more precautions, after his near brush with death.

As Thomas Bearden and others have shown, Scalar EM waves are both gravitational and time waves. Wilbert Smith, an Electrical Engineer and past head of the Department of Comunications in Canada, has experimented with "tempic waves" produced by sending EM pulses through bifilar windings on ferrite cores.

Bifilar windings of copper wire are 2 insulated wires wound

together over the ferrite core. If, on one bifilar coil end, the wires are connected together and a signal is introduced into the 2 wires on the other coil end, the signal current will go in both directions through the coil windings. So, the magnetic field of one wire will "buck" or oppose the magnetic field of the other wire.

In regular electromagnetic theory, there would be a zero resultant magnetic field produced. In extended electromagnetics, a scalar EM field is produced. Wilbert Smith found that a special type of non-electromagnetic signal came off the end of the ferrite core that could penetrate matter like brick walls, conducting steel plates and salt water. He called this signal a "tensor beam."

Clocks placed in front of the tenser beam would slow down or speed up depending on the polarity of the pulses. After discovering these beams could affect time, he called the waves being transmitted, "tempic waves."

Smith had previously experimented with a Tesla interplanetary communications device and apparently was successful in communicating with persons who Smith called the "Boys from the Topside." They were the ones that gave Smith the idea to create his tempic coil. Smith fed the output from a 1 kilowatt transmitter into the coil and observed no standard electromagnetic output. The energy just seemed to disappear to him. The Boys from the Topside informed him that the output from the coil was a six dimensional radio wave which they used for instant communication, power transmission and for pushing and pulling things. They also informed him that the universe was 12 dimensional with four "fabrics" of 3 dimensions each. The basic building block of the universe was spin.

Wilbert Smith used this knowledge and some of his own experiments to create a theory which he called "The New Science" which is available at: http://www.rexresearch.com/smith/news-ci.htm

Smith also determined that space had a "binding energy" that changed from location to location. This binding energy could be disrupted by nuclear explosions for long periods of time. UFOs also disrupted this binding energy. He developed a meter which could measure this binding energy and warned the Aviation industry that planes flying into areas of low binding energy could have their structures weakened to the point of coming apart in midair. Since the science behind this was not understood, his warnings were largely ignored.

Smith was also involved with the secret Project Magnet conducted by Canada to investigate ways to tap the Earth's magnetic field for propulsion and energy.

Thomas Bearden states that the energy from a Smith Coil is a scalar EM wave which oscillates along the time axis and can transmit energy instantaneously since it doesn't traverse space.

Thomas Bearden states in *Energy from the Vacuum*: "When we clearly differentiate the difference between the unobserved (causal) 4-field or dynamic from the iteratively observed 3-field or dynamic, then the non-observed 4-field or dynamic is not limited to light speed. Propagation along the time axis e.g., can be at "infinite velocity" because a single point in time is connected to every point in the universe simultaneously."

The iteratively observed 3-field or dynamic is because each observation in 3-space involves a time derivative (d/dt) operation on 4-space objects. By way of example, the still frames in a moving picture film are time derivatives of the movie.

Bearden then goes on to say: "Absolutely nothing propagates through 3-space! If general relativity is reinterpreted to account for the difference between the observed and unobserved (effect and cause), then the unobserved cause can propagate superluminally without violation of the 'observed event (effect) propagating at light speed' of general relativity. It is the notion that anything propagates in 3-space that is the non sequitur, and should be removed from special and general relativity as well as electrodynamics."

Bearden also says that the electric charge is a 4-flow of energy which enters the electron from the time domain, propagates radially from the electron in 3-space to the proton which brings it in to its center and back into the time domain.

Also, Bearden says that time, like mass can be a source of energy. So to the concept that energy equals mass times the velocity of light squared, we can add that energy equals time times the velocity of light squared so that 1 second = 9 times 10 to the 16th power Joules. The technical problem, of course, is how to convert time into energy.

Bearden states, in a simplification, that the electron rotates 720 degrees in one revolution 360 degrees in real 3-D space and 360 degrees in imaginary (ict) space. During the imaginary part of the electron's spin, it absorbs a small fraction of time energy and during the real portion of the electron's spin, it gives off this energy in a form that creates negative charge properties. This negative charge property flows to the positively charged proton which has a similar 720 degree rotation. This charge as time energy is taken in by the Proton in 3-D space and placed back into the time domain during the imaginary part of its spin creating the 4-flow of energy which is conserved and which appears as an Electric field in 3-space. In fact, every dipole takes some time energy and converts it into EM energy.

Another concept is "time density". A scalar EM wave has no vectorial direction in 3-Space and only oscillates along the time axis. So what exactly oscillates? The answer is the density of time. High density time allows changes to happen easily, low density time causes change to happen with greater difficulty. The Scalar EM wave oscillates about the local average time density. Time density can also be expressed as time rate of flow. Time flows faster with higher time density and slower with lower time density.

In some experiments time charge is not conserved, as with normal 4 – flow between electrons and protons, and a time charge can build up. Bearden states that these time charges explain the anomalous results of some "Cold Fusion" experiments.

There isn't too much in the open literature in the West on the physics of time. That doesn't mean that scientists in the west are not experimenting with time. It just means that time experimentation is mostly being kept secret. In the former Soviet Union, secret experimentation with time was also progressing. After the dissolution of the USSR and a period of "glasnost" (openness) in Russia, a number of scientists who had experimented with and produced theories on the physical nature of time have published their findings.

One pioneering Russian scientist who spent years exploring the nature of time is Nikolay A. Kozyrev. In his theory, time has active properties which affect the cause-effect relationships in physical objects.

In physics to be rigorous, everything has to be causal. That is, first there is a cause and then there is an effect produced by the cause. Kozyrev, through numerous experiments, was able to measure the speed at which cause creates an effect at a distance from the cause. Interestingly, the speed was always the same, even for very different types of experiments.

Even more interesting, this measured speed was the product of the fine structure constant and the speed of light. The fine structure constant is the ratio between an electron's spin momentum and its orbital momentum and is equal to approximately 1/137. So, one property of time is that it allows cause to flow into effect at 1/137 of the speed of light. This is one clue that there is a physical relationship between time and electromagnetism, since light is an electromagnetic phenomenon. This clue is reinforced by Thomas Bearden's concepts on extended electromagnetics.

Analysis of light from distant quasars has indicated that, in the past, the fine structure constant was slightly smaller.

Russian scientist, Dr. Spartak M. Polyakov, worked on the internal structure of the photon (the quantum of electromagnetic radiation). His finding is that the minimum length of the photon is the wavelength of the photon times the ratio between the fine structure constant and the speed of light.

Another interesting fact is that the velocity of the electron in the ground state of Hydrogen is also 1/137 the speed of light. So this fine structure constant seems to have considerable significance! Other Russian scientists have demonstrated that the fine structure constant is a basic property of Aether and is included in a formula calculating planks constant, the quantum of action.

Kozyrev also did some interesting experiments with torsion

fields and waves. Torsion fields are generated by spinning objects. The torsion fields are magnified further when the spinning objects are magnetized and spun around the polar axis of the magnetic field in a direction that reinforces the magnet's torsion field. Any object that spins, including planets and stars give off torsion waves and fields.

Kozyrev developed detectors for torsion fields. Some of these detectors were placed in telescopes aimed at stars. Metal screens were placed in front of the telescope's mirror which blocked normal light but allowed the torsion fields to pass through. A torsion reading could be detected at the star's visible location. This was labeled the past time reading.

However the star's torsion field maximum was not detected at the visual position of the star, which is not where the star actually is, since it takes years for the light from the star to arrive here. The star's maximum torsion field was detected at the calculated position where the star actually was at the time of the experiment! This is pretty significant. It means that torsion fields travel through space very much faster than light! They may even be instantaneous.

Even more interesting was the fact that when the telescope was pointed to the star's future location, that was as far into the future as the past reading was in the past, another reading was detected! Gennedy Shipov would later theorize that torsion fields generated a future and a past wave as well as a present wave.

Kozyrev also states that time exists everywhere simultaneously in the universe. This is a different concept than Einstein's concept of relative time. Kozyrev's experiments have already proven that torsion waves travel at vastly greater speeds than light, perhaps instantaneously, while Einstein claimed the speed of light was the ultimate speed in the universe. So, physical theory must move on past Einstein.

The published literature on torsion fields is significantly greater in Russia than the U.S. Torsion field generators can be purchased from Russia. Experiments with torsion fields have shown that they can also alter properties of materials. So, they must directly interact with the atomic nucleus. Also, Scalar EM waves transport energy instantaneously to remote locations, not through space, but through the time axis. Since scalar EM waves also interact directly with the nucleus, they seem to share some same characteristics with torsion waves.

Kozyrev discovered that spinning gyroscopes do not obey Newtonian principles. The weight of the gyroscope varies depending on the direction of rotation and the angular velocity of rotation. He also found that a constantly spinning object puts out a static torsion field while a nutating or vibrating spinning object puts out a torsion wave.

These torsion waves and fields pass through most objects - but not all. It was discovered that aluminum would shield torsion fields.

Torsion fields may also be the link to consciousness itself. The connection between psychophysical phenomena and torsion field dynamics is discussed in papers titled *Consciousness and the Physical World* produced by Akimov et al.

There is such a vast subject in experiments with torsion fields and waves that I cannot cover the subject adequately. Here is an informative article in English on the subject: http://www.clayandiron.com/news.jhtml?method=views&news.id=1509

The world's leading physicist on torsion fields is the Russian Dr. Gennedy Shipov who has written the book *Theory of Physical Vacuum*. He has the following to say about the instantaneous propagation speed of torsion fields:

"In the *Theory of Physical Vacuum*, the Primary Torsion Fields there in the vacuum are the space-time vortexes with zero energy and with interaction without energy. For an object, where the energy is equal to zero, it is impossible to formulate a concept of the speed of its propagation. For the usual observer such an object is "at once everywhere and always", i.e. its "speed of propagation" is instantaneous. The creation of primary torsion fields can be considered as a primary polarization of the vacuum according to its spin properties, with right-hand and left-hand fields arising simultaneously. Experiments on creation of artificial torsion polarization of the physical vacuum would introduce in some of its area material objects with various surface geometry, that show right-hand and left-hand primary torsion fields that arise simultaneously. The geometry of space in this case represents a 10-dimensional manifold (4 translational coordinates and 6 angular ones), and its Riemannian curvature are equal to zero, and the Ricci torsion being distinctively different from zero. Propagation of primary torsion fields with "instantaneous speeds" happens on phase portrait of these fields, but not with the help of looking at group velocity, as it occurs with usual physical fields. This phenomenon indicates a holographic structure of torsion fields."

Shipov also had this to say about the possibility to unify Relativity theory with Quantum theory:

"Recently I have written an article entitled "Rotational Relativity and Quantum Mechanics". It will soon be translated into the English language. In this article, I show in detail that Ricci torsion plays a role of the bridge between the theory of Relativity and the Quantum theory of a matter. Apparently that mysterious wave function Y of the New Quantum theory = a Matter field = Ricci torsion field. As a Ricci torsion field describes fields of inertia, the quantum mechanics is a way to describe movement of matter through the dynamics of its inertial field. I think that after a while all abnormal consequences and contradictions of quantum mechanics will be interpreted in a clear language of torsion fields - new physical object. Einstein dreamed of this."

Gennedy Shipov has also created the start-up UVITOR Company which has made some breakthrough experiments in Vacuum Metric Teleportation using a 4 dimensional gyroscope or 4-D Warp engine. This is the first step towards topological teleportation, where the object falls into the wormhole and appears at a different space-time point. Their Web site is: http://shipov.com

Thomas Bearden, in his book *Energy from the Vacuum: Concepts and Principles,* states that scalar EM is given off by all astronomical objects like galaxies, stars, planets and satellites. Also, nucleons exchange scalar radiation. He also claims that the property of mass and inertia is caused by trapped scalar EM radiation and the means of trapping is the particles' spin. The spin factor further suggests that scalar EM and torsion fields may definitely be related.

In the November – December 2001 issue of *New Energy Technologies* magazine (Issue #3), published by Faraday Lab Ltd., http://www.faraday.ru/ an experimental time machine is featured. The time machine was designed and built by the private Russian research association named Kosmopoisk and headed by Dr. Vadim Chernobrov.

So as not to sound too sensational, their time machine was only able to speed up or slow down the course of time by a small percentage (around 3%) of its normal flow rate. No trips far into the past or future were made. However from an experimental time research project point of view, much was learned about engineering time from their experiments.

Basically, their time machine was made from 3 concentric fiberglass spheres with the center volume, the test region. Each sphere had numerous special electromagnetic solenoids embedded in the surface with the solenoid axis radially directed towards the center test region. These solenoids were simultaneously pulsed with electric current so that the magnetic pulses all converged to the center of the sphere. Even more detailed information on this time machine was in the 2005 issue # 23 of *New Energy Technologies* magazine, which illustrated the unique construction of the solenoid coils.

Their time machine was first tested on mice. At first, some of them died. The problem was caused because different parts of their bodies were experiencing different time flow rates. Also the temperature was too high in the test region. After these problems were solved and no more mice were dying, they selected a stray dog for a test subject. No harmful effects were discovered on the dog. So, they enlarged the time machine and started looking for human volunteers.

They tested men and women. The women had higher emotional experiences than the men did. One woman experienced astral projection. Also, speeding up time had different effects than slowing time down had. Effects were more erratic and unstable when speeding time up. In this regime, outer influences like tidal effects of the sun and moon had greater effects on what went on inside the time machine.

Human observers inside the time machine could see areas of space having different time rates. These time rate boundaries had the appearance of a white mist. The higher the difference of time rate, the denser the white mist would appear. That could give a visual warning of a biological hazard area to avoid. Also, strange lighting effects would appear in the sky over the time machine when it was in operation.

Since all the electromagnetic solenoids in this time machine were directing their magnetic fields towards the center, the magnetic fields vectorialy cancel in the center, creating a scalar electromagnetic field in that area.

So, here is more experimental proof that macroscopic time is affected by scalar EM fields. The experimental alteration of the rate of time flow by scalar electromagnetic fields opens up a whole new area of fundamental physics.

Another scientist from Belarus, Albert Veinik followed up on Dr. Vadim Chernobrov's time machine. Veinik ran experiments on flywheels in which different time flow rates acted on different parts of the flywheel. The results of this experiment showed that the centrifugal force was asymmetrical on the flywheel.

This effect can be used as another means of propulsion on the planet and in space. Suppose you generated a positive scalar field on one half of a flywheel and a negative scalar field on the other side. The positive scalar field would increase the time flow rate on one half of the fly wheel causing that half to generate a greater centrifugal force. The other half of the flywheel exposed to the negative scalar field would have a lower time rate of flow and a lesser centrifugal force generated. This would result in an unbalanced force that could be used for propulsion.

Veinik also developed the concept of a time charge of Cronons which were the quantums of time. He theorized that every object has an intrinsic Cronal charge which is caused by the number of Cronons present in the object. This Coronal Charge probably is similar to Thomas Bearden's time charge concept.

In his book, *UFOs Explained at Last, The Science of Extraterrestrials,* Eric Julien proposes that time actually has 3 dimensions; density, direction and present. This concept is used by Eric Julien to unify quantum physics with relativistic physics in his book. Eric Julien, like Thomas Bearden, also says that time is energy.

Dr. David Lewis Anderson also has claimed to have experimented with time control technology. He states that there is a secret "time race" around the world among countries like the U.S., India, Russia, China and Japan to develop better time control machines.

Dr. Anderson has exposed human beings to retarded and accelerated rates of time. Persons exposed to this change of time rate experience deep emotional states and expanded consciousness and feelings of spirituality. Dr. Anderson has not disclosed the exact rate of acceleration or retardation of time in his device.

Dr. Anderson's web site is here: http://andersoninstitue.com

4

German Flying Saucers

Maria Orsic was a famous medium in Germany. She was born in Vienna, Austria on October 10, 1895. Her father, Tomislav Orsic, was a Croatian immigrant from Zagreb, her mother, Sabine Orsic, was from Vienna. Later, Maria Orsic moved to Munich, Germany.

In Munich, Maria was in contact with the Thule Gesellschaft and soon she created her own circle together with Traute A. from Munich and several other friends called the Alldeutsche Gesellschaft für Metaphysik (All German Society for Metaphysics). Later, the name was changed to Vril Gesellschaft (Vril Society).

The members were all young ladies. Both Maria and Traute were beautiful ladies with very long hair; Maria was blond and Traute was brown-haired. They had long pony tails, a very uncommon hairstyle at that time. This became a distinctive characteristic in all the women who joined the Vril Society, which was maintained till May 1945. They believed that their long hair acted as cosmic antennae to receive alien communication from beyond.

On February 10, 1917 Maria Orsic, fell into a trance or coma that lasted several hours, in which she claimed that beings of light from another world had communicated with her. She had reoccurring incidences of this communication in the days that followed. She was told not to tell anyone about her communication with them except the mediums Traute, Gudrun, Sigurn and Heike.

She visited her medium friends and explained her experience to them. They were quite understanding and even told Maria that they were expecting her to come to them. Later, the beings of light informed Maria that they were from the Star system they called Aldebaran (known as Alpha Tauri in the Taurus Constellation to our astronomers).

The telepathic communications with Maria Orsic contained two types of information; (1) metaphysical revelations, history of ancient civilizations on our planet and the true origins of the human race, (2) technical information on how to build a super flying machine. Also, information in an unknown script was given through trance automatic writing, which remained to be translated. The script was given to the "Panbabylonists", a circle close to the Thule Society which was integrated by Hugo Winckler, Peter Jensen, Friedrich Delitzsch and others. It turned out that the mysterious language was actually ancient Sumerian. Sigrun, from the Vril gesellschaft helped translate the language and decipher the strange mental images of a circular flight machine.

Maria Orsic could not understand the technical information and asked her father, Tomislav, to help her understand what it meant. He also was baffled. However, he was acquainted with Professor Winfried Otto Schumann, a scientist, who taught at the Technical University of Munich. After Mr. Orsic conveyed Maria's information to Dr. Schumann, he became quite interested.

Dr. Schumann was already familiar with Victor Shauberger's work with spiraling air and water implosion technology leading to levitation and Dr. Schapeller's work with "glowing magnetism" and extracting energy from the Aether. Therefore, he thought the information relayed by the Orsics, which didn't contradict the other's work, could have potential.

The concept of "other science" (or "alternative science") matured during this time and the following years. Because of the financing difficulties it took three years until the flying machine project started taking shape. Both the Vril and the Thule society members contributed to the effort. By 1922, parts for the machine began arriving independently from various industrial sources paid in full by Thule and Vril. A team of engineers and investors was organized to build the super flying machine which they dubbed the Jenseitsflugmaschine (JFM) (otherworld flying machine).

However, it should be important to consider the possible motivation behind the Aldebaran civilization's offer to assist the Vril Group and Germany.

Researcher Wendelle Stevens tells us that, rather than a militant gesture of aid to aggressive Nazis, the Aldebarans perceived an economic disparity in Earth cultures that fueled perpetual wars and conflict.

To alleviate this disparity the Aldebarans reasoned that by offering 'free-energy' technologies, used to create affordable mass transportation devices, a new innovative generation of industries, promoting prosperity and greater peaceful interaction between nations might result; thus diminishing violent wars.

Clearly such a plan resonated with members of both the Thule and Vril Societies and their dream for a utopian New World based on 'alternative science'.

Upon studying these otherworldly, esoteric designs, Dr. W.O. Schumann and his associates from the University of Munich realized the channeling actually contained viable physics, and over the ensuing years construction was initiated to make this flying machine a reality.

By 1922 development of a working prototype was underway. Meanwhile, Germany saw the inception of the National Socialist Party and Adolph Hitler's rise to power, fueled in part by the utopian visions of a new world order inspired in part by the Thule and Vril Societies.

On March 22, 1922, the first model was tested and failed miserably. It rose about 50 feet into the air. It spun around like a giant pinwheel spouting fire and disintegrated. The pilot barely escaped with his life. They went back to the drawing boards and Maria Orsic went into another trance to get further guidance from the light beings. Several days later, Maria returned to Schumann giving him some new notes and pictures on how to rectify the problem. A new twist was that Maria informed Schumann that mental control had to be used to fly the machine. Schumann was ready to throw in the towel.

Later, he continued on with the project when Maria presented him with the complete plans and instructions for a "Head Band Mental Command Device." As events progressed, Dr. Schumann not only became a father figure for Maria but also is considered the father of the German Flying Saucers.

By December 17, 1923, the second model of the Jenseitsflugmaschine (JFM2) was tested. During this time Maria and

Sigrun made 8 visits to the hangar where the craft was located and gave their channeled findings to the engineers before the test.

This remote controlled test was quite successful; the craft flew at quite a high speed for 55 minutes. However after the craft landed, it looked very weather worn instead of brand new. Maria explained this was because the craft was flying in a parallel dimension that caused alterations in the materials of the craft and could have the same effects on people flying in the craft. This explanation horrified the engineers present. Dr. Schumann was not, however, dismayed because he had Maria Orsic, who could tap into the metaphysical world to solve any problems.

The JFM2 levitator unit was further developed by Schumann and others into RFZ (Rundflugzeug) class of flying saucers in 1934 and the Vril and Haunebu disks of 1939-1945. Dr. Schumann is considered the inventor of the Schumann Levitation Disk.

In late November 1924, Maria Orsic visited Rudolf Hess in his apartment in Munich, together with Rudolf von Sebottendorf, the founder of the Thule Gesellschaft. Sebottendorf wanted to contact Dietrich Eckart, who was a famous former Thule member that had deceased one year before. To establish contact with Eckart, Sebottendorff and other Thulists (amongst them Ernst Schulte-Strathauss) joined hands around a black-draped table.

Hess found it unnerving to watch Maria Orsic's eyeballs rolling back and showing only whites, and to see her slumping backward in her chair, mouth agape. However Sebottendorff smiled in satisfaction as the voice of Eckart started coming out of the medium. Eckart announced that he was obliged to let someone else's voice come through, with an important message. A weird voice then identified itself as "the Sumi, dwellers of a distant world, which orbits the star Aldebaran in the constellation you call Taurus the Bull".

Hess and Schulte-Strathaus blinked at each other in surprise. According to the voice, the Sumi were an humanoid race who had briefly colonized Earth 500 million years ago. The ruins of ancient Larsa, Shurrupak and Nippur in Iraq had been built by them. Those of them who survived the great flood of Utnapishtim (the Deluge of Noah's Ark) had become the ancestors of the Aryan race.

Sebottendorff remained skeptical and asked for proof. While Maria was still in a trance, she scribbled several lines of queer-looking marks. Those marks turned out to be ancient Summerian characters, the language of the founders of the oldest Babylonian culture.

The Vril mediums had received precise information regarding the habitable planets around the sun Aldebaran and they were willing to plan a trip there. This project was discussed again the 22nd January 1944 in a meeting between Hitler, Himmler, Dr. W. Schumann and Kunkel of the Vril Gesellschaft. It was decided that a Vril 7 "Jäger" would be sent through a dimension channel independent of the speed of light to Aldebaran.

According to N. Ratthofer (a writer), a first test flight in the dimension channel took place in late 1944. The test flight almost ended in disaster because after the flight, the Vril 7 looked "as if it had been flying for a hundred years". Its outer skin looked aged and had suffered damages in several places. This seemed to be a hazard of extra dimensional travel.

Another improved type of saucer developed from information from Maria Orsic and the Vril Society was the Vril 7 Giest which was built at Arado-Brandenburg and flew in 1944. Maria was able to fly this model telepathically without the Mind Control Head Band built earlier. Maria was spiritually evolved and did not wish for the Vril 7 Giest to be used for war. So, she stated that further improvements were needed and the saucer was sent to a hangar in Munich for further improvements.

She also had two smaller saucers of 27 foot diameter built. Dr. Schumann recruited 4 engineers to build these craft. She was secretly planning to escape Germany and join her Aldebaran friends who had already informed her that Germany would lose the War.

In 1944, Heinrich Himmler, head of the SS, replaced Albert Speer's appointee, Georg Klein, with Dr. Hans Kammler as overseer of this combined saucer project. Himmler's employee, Kammler, replaced Speer while Klein did what he always did. The result was that the SS took direct and absolute control over these projects from this point until the end of the war.

Prior to this happening, news of these designs or application itself was made to the German Patent Office. All German wartime patents were carried off as booty by the Allies after the war. This amounted to truckloads of information. Fortunately, Rudolf Lusar, an engineer who worked in the German Patent Office during this time period, wrote *German Secret Weapons of the Second World War* in the 1950s listing and describing some of the more interesting patents and processes based upon his memory of them. They are surprisingly detailed. Included is the Schriever saucer design with detail. Also discussed are the Miethe Belluzzo saucer designs.

In the early stages, these different groups worked independently. But, as the War drug on, and the saucer projects were brought under the command of the SS, more information was shared between the groups to further improve the German flying saucers.

Saucers based on Maria Orsic's information were not the only types being built in Germany. Victor Schauberger had also independently developed his "Repulsine" which would develop diamagnetic levitation forces.

The Repulsine consisted of a rotating chamber which had sine wave ripples in the radial direction. The side of the chamber ripples facing the center had intake slots cut in to allow the intake of fluids, either air or water. Centrifugal force would force the fluid to the outside of the chamber and further down the sine wave shaped chamber to the periphery where the fluid would strike vanes that would force the fluid downward.

The fluid would be spun so fast that its velocity exceeded the speed of sound in that fluid. Just the fluid dynamics would create tremendous vertical forces. These forces were enhanced further by levitational forces that Schauberger claimed that his Impulsine was generating.

Also the fluid striking the vanes would cause rotation of the device in excess of what the electric drive motor would cause. So, after a certain amount of RPMs, the device would power itself. The original idea was to use the Impulsine, under water, in the nose of a submarine where the fluid would be ejected rearward creating vortices along the tapered rear of the submarine. These vortexes would reduce friction while causing pressure on the hull pushing it forward. This concept, Schauberger called his Trout Turbine since he got the idea from studying trout swimming up waterfalls.

At first, the SS was interested in using Victor Schauberger's

"Trout Turbine" to propel submarines. At Mauthausen, under orders from Heinrich Himmler himself, Schauberger was to carry out research and development for the Third Reich war effort. He was given approximately 20-30 prisoner engineers to proceed with his research into what was termed "higher atomic energies".

Development of Shauberger's discoid motor continued until one of the early test models was ready for a laboratory test that ended in disaster. The model was 2.4 meters in diameter with a small high-speed electric motor. The whole unit was securely bolted down with heavy bolts embedded into the concrete floor. Upon initial start-up the Repulsin A was set in motion violently and rose vertically, quickly hitting the ceiling of the laboratory, shattering to pieces. Obviously the forces generated by the device were greatly under estimated and the unit had sheared the mounting bolts. The SS were not pleased and even threatened Schauberger's life, suspecting deliberate sabotage.

Replacement models were built, but by 1943 a more improved design, the Repulsin B model was constructed with the SS objective of developing this motor for an odd SS bio-submarine which Schauberger named the "Forelle" (Trout) due to its configuration of a fish with a gaping mouth!

Finally, it appears that the SS had discarded the idea of applying the Schauberger motor to a submarine when the benefits would greatly improve their work on the secret Flugkreisel (flying top) which was taken from Rudolf Schriever back in 1941.

In June 1934, Viktor Schauberger was invited by Hitler and the highest representatives of the Vril and Thule societies and worked, from then on, with them. The first German UFO was developed in June 1934. Under the direction of Dr. W. O.

Schumann, the experimental circular aircraft RFZ 1 was developed at the Arado aircraft factory in Brandenburg.

Even though its first flight was also its last, it rose vertically to a height of about 60 meters but then started to tumble and dance in the air for several minutes.

The Arado 196 tail unit which was supposed to guide the device proved to be completely ineffective. With great difficulty the pilot Lothar Waiz succeeded in bringing it back to the ground, escaping from it, and getting away in time, because the device started to spin like a top before it overturned and completely broke into pieces. That was the end of the RFZ 1 but the beginning of the VRIL flying machines.

The RFZ 2 was finished at the end of 1934; it had a Vril drive and a magnetic impulsion flying system. Its diameter was 5 meters and had the following characteristics: the contours of the device became blurred as it gained speed, and it lit up with different colors, a well-known characteristic of UFOs.

Depending on the propulsive force, it became red, orange, yellow, green, white, blue or violet. It was able to operate, and it had a remarkable destiny in the year 1941. It was used as a long range reconnaissance aircraft during the battle of England.

The RFZ 2 also was photographed at the end of 1941 over the south Atlantic while on the way to the auxiliary cruiser *Atlantis*, which was in Antarctic waters.

After the success of the small RFZ 2 as a long range reconnaissance aircraft, the Vril Society acquired its own testing ground in Brandenburg. The VRIL 1 Hunter, a lightly armed flying disc, flew at the end of 1942. It was 11.5 meters in diameter, had a single seat, and it had a Schumann levitation drive and a magnetic field impulsion flying system.

The Schumann levitation drive had two counter-rotating disks inside which created two counter rotating torsion fields. Counter-rotating magnetic fields of the correct polarity could also magnify these torsion fields.

As opposing electromagnetic fields create scalar EM fields which cease to exist in 3-D space and work in the time dimension, counter-rotating torsion fields of sufficient strength cause objects inside the influence of these fields to become isolated or shielded from surrounding space–time.

Under these conditions, gravitational and inertial forces have no influence on the object in the shielded space-time and the object can easily levitate and even undergo tremendous acceleration with no force felt by the object.

The Schumann levitator, based on channeled Aldebaran technology, is totally different than Schauberger's levitation devices and Tom Townsend Brown's electrogravitics. However, the Germans, working under Hans Kammler, probably incorporated ideas from all three inventors to evolve the most efficient craft for the various tasks at hand.

Information and pictures of Schauberger's Repulsine are in the book *The Energy Evolution* by Callum Coats and at: http://discaircraft.greyfalcon.us/Viktor%20Schauberger.htm/_

Even though he doesn't believe in the extraterrestrial presence, Henry Stevens' book, *Hitler's Flying Saucers* is also a good source of information on the flying saucers of the Third Reich.

The Italians were also building flying saucers jointly with the Germans. Belluzzo and Miethe teamed up to build a flying saucer that was later described in the Italian press. Rudolf Schriver had also designed a saucer and joined the Belluzzo and Miethe team. Schriver was also a test pilot of these designs.

A joint saucer design with a diameter of some 45 meters, by the specialists Schriever, Habermohl and Miethe was perfected. They were first airborne on February 14, 1945, over Prague and reached a height of nearly eight miles in three minutes. They had a flying speed of 1,250 M.P.H. which was doubled in subsequent tests. It is believed that after the war Habermohl fell into the hands of the Russians. Miethe developed at a later date, similar flying saucers at A. V. Roe and Company for the United States.

The SS E-IV (Entwicklungsstelle 4), a development unit of the SS occult "Order of the Black Sun" was tasked with researching alternative energies. By 1939, this group developed a revolutionary electro-magnetic-gravitic engine which improved Hans Coler's free energy machine into an energy converter coupled to a Van De Graff band generator and Marconi vortex dynamo (a spherical tank of mercury) to create powerful rotating electromagnetic fields that affected gravity and reduced mass. It was designated the Thule Triebwerk (Thrust work) drive and was to be installed into a Thule designed disc.

The Hans Coler free energy device was invented by German naval Captain Hans Coler in the 1920s. His device was examined by Dr. M. Close of the Berlin School of Technology and Dr. W.O. Schumann and verified to work in 1926. In 1942, he improved his "current generator" into a device which produced a continuous output of 7 kilowatts with a battery input of 8 watts and powered his home for the last 3 years of the war with his device.

The German General Headquarters Naval Research Section examined Coler's Device on April 1, 1943. The Coler device was placed into production by the Continental Metal AG, Reinmetall-Borsig, Siemens-Schuckert, and Hermann-Goring-Werke companies in contracts with the German Navy. This production ceased at the end of the war and never has been reestablished.

Since 1935 Thule had been scouting for a remote, inconspicuous, underdeveloped testing ground for their Thule designed craft. Thule found a location in Northwest Germany that was known as Hauneburg. At the establishment of this testing ground and facilities the SS E-IV unit simply referred to the new Thule disk as Haunebu in 1939 and briefly designated it the RFZ-5 along with Vril's other machines which included the Rundflugzeug RFZ-1 through the RFZ-4.

At a much later time in the war, as production of these craft was to commence, the Hauneburg site was abandoned in favor of the more suitable Vril Arado Brandenburg aircraft testing grounds. Although designated as part of the RFZ series the Haunebu disc was actually a separate Thule product constructed with the help of the SS E-IV unit while the RFZ series were primarily built at Arado Brandenburg under Vril direction up to the RFZ-4 disc. Three models were built, the Haunebu I, Haunebu II and Haunebu III.

The early Haunebu I craft of which two prototypes were constructed were 25 meters in diameter, had a crew of eight and could achieve the incredible initial velocity of 4,800 km/h, but at low altitude. Further enhancement enabled the machine to reach 17,000 km/h! Flight endurance was 18 hours. To resist the incredible temperatures of these velocities a special armor called Victalen was pioneered by SS metallurgists specifically for both the Haunebu and Vril series of disc craft. The Haunebu I first flew in 1939 and both prototypes made 52 test flights.

In 1942, the enlarged Haunebu II of 26 meters diameter was ready for flight testing. This disc had a crew of nine and could also achieve supersonic flight of between 6,000-21,000 km/h with a flight endurance of 55 hours. Both it and the further developed 32 meter diameter Do-Stra had heat shielding of two hulls of Victalen. Seven of these craft were constructed and tested between 1943-44. The craft made 106 test flights.

By 1944, the perfected war model, the Haunebu II Do-Stra (Dornier STRAtosphären Flugzeug) was tested. Two prototypes were built. These massive machines, several stories tall, were crewed by 20 men. They were also capable of hypersonic speed beyond 21,000 km/h. The close of the war, however, prevented Dornier from building any production models.

The Haunebu III was a larger 71 meter diameter. A lone prototype was constructed before the close of the war. It was crewed by 32 and could achieve speeds of between 7,000 - 40,000 km/h! It had a triple Victalen hull. It is said to have had a flight endurance of between 7-8 weeks! The craft made 19 test flights and was likely used to evacuate the Thule and Vril people along with Hans Kammler at the end of the war. Both societies disappeared at the close of the war.

Most of these craft were under development at the time of Germany's defeat and weren't used in actual combat with a few exceptions. In 1944 a massive bombing raid was launched against the critical ball-bearing plant at Schweinfurt. Within a matter of hours a squadron of ten to fifteen Nazi discs managed to shoot down as many as one-hundred and fifty British and American bombers - one quarter of the entire bomber contingent.

Still, facing overwhelming odds, the crumbling Reich lacked sufficient saucer squadrons to turn the tide. If Germany had waited until these wonder weapons were fully operational before starting their military adventures, they would have most likely have won the war. Lucky for us, their leaders were too impulsive to wait!

At the war's end, Hans Kammler directed an evacuation of advanced Nazi SS technology and paperwork using a Special Evacuation *Kommando* organized by Martin Bormann. JU 290s and a JU 390 "flying trucks" with tremendous range were used to evacuate the Nazi VIPs, scientists, and technical material to Argentina. Many of the German Saucers were flown to secret Nazi bases in Argentina and Nue Schabenland, Antarctica.

It is estimated that about 100,000 Nazis escaped to Argentina with considerable Nazi gold after the war. This operation was organized by Organisation Der Ehemaligen SS-Angehorign (ODESSA, meaning Organization for Former SS Members.)

Considerable land was purchased by the Germans before the War and by ODESSA after the war around the San Carlos de Bariloche region that, in total, measured 160 Kilometers by 170 kilometers. Many underground bases and laboratories and above ground roads, bridges and an airport were constructed there.

Above ground, the area has all the appearance of a typical European Alpine community and these days is a favorite resort area.

In the mid 1950's Hans Kammler was seen in the company Bormann, Otto Skorzney, Reinhard Gehlen and German industrialists at a resort meeting at San Carlos de Bariloche, Argentina.

The Nue Schabenland base in Antarctica (also called Queen Maud Land), was established in 1938 when a German expedition discovered huge ice caves tunneled out by a warm river that was geothermally heated. These ice tunnels made perfect hiding places for submarines under the Antarctic ice cap and a large secret underground base called Nue Berlin was constructed in the area.

After the War, England had intelligence on the German secret Nue Schabenland base and suspected that many of the unaccounted for German U-boats might be hiding there. In the later months of 1945, British elite forces were being trained on South Georgia Island in Antarctic warfare.

After months of rigorous training these forces were sent to the British base in Maudheim, Antarctica. Their first mission was to investigate anomalous activities around the Mühlig-Hoffmann Mountains. A tunnel was discovered in dry, snow free valley. A force was sent to explore the tunnel which went a long way. At the end of the tunnel was a huge cave which was partly filled with sea water.

Then, the exploratory team discovered the U-Boat sub pens and hangars for strange looking aircraft built right into the rock of the cavern. The place was crawling with Nazi soldiers and technicians. The base was so huge that it looked like the Germans had been working on it for a long time. The elite forces decided to sabotage the base as best they could. They were packing plenty of explosives and started stealthily planting mines in strategic locations, which would cause rock falls on strategic parts of the base, and targeted fuel and ammo dumps.

They were eventually spotted by the Germans and had to flee back out the tunnel with the Germans chasing and firing after them. Of the thirty men that went in, only ten came out. They blew up the tunnel entrance blocking it to the Germans and the ten were evacuated back to Maudheim base and then back to the Falkland Islands. All the survivors were sworn to secrecy and the whole event was covered up by the British Government. (5)

In 1947, a U.S. military expedition of an aircraft carrier, a submarine, 20 surface ships and 4,000 elite Navy troops, disguised as a scientific exploration, was launched against this base under the code name "Operation Highjump" led by Admiral Byrd. After eight weeks, Operation Highjump was defeated by German flying saucers and Byrd beat a hasty retreat back to the U.S. Much of Operation Highjump is still classified today.

At the wars end in Germany, any remaining German flying saucers were destroyed and their engineers and scientists killed before the Allies could overrun their bases and gain their technology.

However, the Allies, the U.S., Russia, Briton and Canada did get some of the technical data, partially destroyed hardware and engineers and scientists of the flying saucer program and are keeping it all top secret. The U.S. received the lion's share of the information. Many of the Germans who worked on these projects, including Dr. Schumann, the father of the German flying saucers, were brought to the U.S. in "Operation Paperclip." All in all about 5,000 Paperclip Nazis were brought in. Later, they were hired by the military and defense contractors in their own "black projects."

As for Maria Orsic – she totally disappeared!

On March 14, 1945 Maria Orsic had a farewell meeting with Dr. Schumann, who she loved very dearly. She hugged him with tears in her eyes and said that she regretted that their space ship had not been developed much earlier, for she could have used it to stop the war, alter the fate of Germany and bring peace on Earth.

Dr. Schumann asked her: "And now where to Maria?"

"To the station, where they are waiting for me."

"And will you be flying the Jenseitsflugmaschine…No, I am not going to ask you where did you learn to fly such a monster."

"I will be guided…Did you forget the band you created for me?"

"How far will it take you?"

"To the station, and from there, their mother ship will take me to them…I have already explained everything to Traute, Heike and the others and they are happy."

"Will I ever see you again?"

"Not in this world…I will always remember you."

She kissed Dr. Schumann on the fore head and they parted company.

On March 17, 1945 Maria and her Vril group went to the hangars of Messerschmitt in Augsburg, to take possession of the Vril spaceship.

On March 18, 1945, a strange circular shaped flying craft was spotted by Lt. Col. Walter Fellenz, Brigadier General, Henning Linden and others of the U.S. 7th Army which was then advancing on Munich. The craft hovered over Munich and then swiftly disappeared! (6)

5

The Extraterrestrial Presence

Extraterrestrials have been visiting our planet for thousands, perhaps millions of years. The Popol Vuh states, "Men came from the stars, knowing everything and they examined the four corners of the sky and the Earth's round surface." Many researchers, like Zecharia Sitchin, Eric Van Daniken and others have presented a good deal of evidence of past extraterrestrial presence. Many contactees have related their contacts with extraterrestrials in the present age. Since the 1950s, this matter has been purposely hushed up and debunked by the U.S. and other governments.

The first meeting, in recent history, between extraterrestrials and government leaders took place on July 11, 1934 on board a U.S. Naval vessel at the port of Balboa, Panama. The meeting was between the Small greys from the Orion star system and U.S. President Franklin D. Roosevelt. These Greys were actually representing, and controlled by, the Reptilians from the Alpha Draconis star system and definitely did not have the best interests of humans at heart.

Roosevelt was more interested in obtaining extraterrestrial technology than keeping his oath to uphold the Constitution. He entered into a secret (and therefor unlawful, since it was never ratified by the Senate) treaty with the Small Greys to allow them to proceed unhindered with human abductions for use in their genetics program in return for their high technology.

Earlier in 1934, the Pleiadians also attempted to make a deal with president Roosevelt. They would offer economic and social programs to help transform the earth into a paradise. The only condition was that the U.S. would have to give up war and embark on a disarmament program. Roosevelt refused. After that, the Pleiadians approached Hitler and agreed to give him some technology if he promised not to attack the Jewish people. The Nazis got some technology but didn't honor their part of the bargain and the Pleiadians pulled out of the treaty in 1941.

After that, the Greys approached the German government of Adolf Hitler. They entered into a similar treaty that they had made with Roosevelt. But, Hitler got a better agreement; the Greys were not to abduct the German Aryan people. Only people in the concentration camps could be abducted. The treaty of 1934 was renewed in 1944. Only a very few people, probably less than 20 in Germany, England and the US combined, knew about these secret treaties which were renewed every 10 years.

The Small Greys from Orion were different than the Greys that were from the Zeta Reticulum star system, although they were similar in appearance.

Several months after Pearl Harbor, in the early morning hours of February 25, 1942, an unidentified saucer shaped aircraft flew over L.A. County that sparked an air raid alert. Searchlights and Guns were locked onto the aircraft for about an hour as anti-air craft guns fired away at the craft which hovered over the city of Los Angeles. The saucer was unharmed from all the flack shells exploding around it and eventually flew away. The only damage to the city was from falling debris from the anti-aircraft fire. This was reported as the "Battle of Los Angeles" in the Los Angeles Times and other newspapers

What was not well known was that the Navy had recovered a crashed flying saucer about the same time, west of San Diego. Two dead Zeta Reticulans (they didn't know until later where they were from) were found inside the saucer. The craft and alien bodies were taken to the Foreign Technology Section at Wright Patterson Air Force base in Dayton Ohio and studied by the Retfours Special Studies Group.

A Top Secret memo dated March 5, 1942, addressed to President Roosevelt explained how the Navy had recovered an "unidentified airplane... of interplanetary origin". Army Chief of Staff, George Marshall ordered the crashed saucer studied and created a highly classified unit called the Interplanetary Phenomenon Unit (IPU), within the Army's G-2 Intelligence Agency, to do the study.

Undersecretary of the Navy, James Forrestal and General Eisenhower were also informed of the saucer recovery by the Navy. Later, John F. Kennedy, who spent the early part of the war in Naval Intelligence, and was a close friend of Forrestal, would accompany Secretary James Forrestal on a 1945 trip to Europe. At this time, Forrestal informed Kennedy of the recovered saucer and told him to keep it secret.

According to the Army's Directorate of Intelligence, The IPU existed until the 1950s when it was disbanded and all of IPU's records were turned over to the U.S. Air Force Office of Special Investigations in conjunction with Operation Bluebook.

In 1947, a Flying Saucer crashed at Roswell New Mexico. Actually, there were two separate UFO crashes near Roswell as well as others at Corona, New Mexico around the same time.

There were two Corona crash sites. One southwest of Corona, New Mexico and the second site at Pelona Peak, south of Datil, New Mexico. The crash involved two extraterrestrial aircraft. The Corona site was found a day later by an archaeology team. This team reported the crash site to the Lincoln County Sheriff's department. A deputy arrived the next day and summoned a state police officer. One live entity [EBE1] was found hiding behind a rock. The entity was given water but declined food. The entity was later transferred to Los Alamos.

Early in the first week of July, 1947, William Ware "Mack" Brazel was riding in his pickup truck with the son of the Procter family on the J.B. Foster ranch located southeast of Corona and about 75 miles northwest of Roswell and discovered a debris field about ¾ a mile long from a crashed object. Brazel took some of the debris back to his home and showed the pieces to the Proctors. The Proctors told Brazel that the debris could be from an Alien Craft since many UFOs had been spotted in the area recently.

Several days later, Mack Brazel rode into Roswell, the county seat and reported his findings to the Chaves County Sheriff, George Wilcox, who then informed Air Force Major, Jesse Marcell, intelligence officer for the 509th Bomber group stationed at Roswell Army Air Field, which was the nuclear arms wing of the Army Air Force.

The Army Air Force was more concerned about the Roswell Crashes because they wished to limit public knowledge of the UFOs. One UFO crash was just north of the city of Roswell and easily accessible to the public. So, the psychological warfare department of Army's G-2 had to do a "bait and switch operation" to direct attention elsewhere.

On orders from Col. Blanchard, Lt. Walter Haut RAAF public information officer, on July, 8, 1947, leaked a story to the press that the Roswell Army Air Force had recovered a crashed flying saucer on the J.B. Foster Ranch, which was in a more remote location and more difficult for investigators to reach. The Crashed UFO north of the city was quickly recovered and the site cleaned up by special crashed UFO recovery Delta teams, while public attention was directed elsewhere.

Then, Air Force Major Jesse Marcell, posed by some silvery fragments of an alleged balloon while photographs were taken. Meanwhile, a second recovery team was cleaning up the J.B. Foster Ranch crash site. The military turned away any persons or news reporters that approached that recovery site. All the debris was taken to the Roswell Army Air Force base and crated and stored in a hanger and was later flown to Wright Field in Dayton Ohio according to AAF pilot "Pappy" Henderson.

The next day, on July 9, the Air Force's Story was changed to recovered balloon fragments – not a flying saucer. Mack Brazel who had been detained by the military, was escorted to the Roswell Daily Record by three military officers and changed his original story. Mack Brazel was finally released from military custody on July 12. After that the story quickly died out.

Some local people, however, knew different. The Brazels and Procters had seen the actual debris from the crash site which was later confiscated by the military. Some I-beam shaped material was as light as balsa wood, flexible and tougher than steel. Many strange hieroglyphics were inscribed in the I-beams in violet color. They had been threatened by the military not to talk about the incident, if they didn't want their families hurt.

The ET bodies from the UFO crashes were packed in dry ice taken from Roswell train station and first taken to Los Alamos National Laboratory because they had a freezing system that allowed the bodies to remain frozen for research. All the remains were eventually taken to Wright Paterson Air Force Base (then called Wright Army Air Field) in Daton, Ohio. Wright Paterson was chosen because it already had the facilities and personnel to analyze and back engineer recovered alien technology.

All of the UFO recoveries were reported to President Truman who then signed a memo to then Secretary of Defense, James Forrestal, authorizing the creation of "Operation Majestic Twelve" headed by Dr. Vannevar Bush, then director of the Office of Scientific Research and Development.

MJ-12 or Majestic-12 would be a secret group of two leaders from each military service; Air Force, Navy and Army and six leaders in the scientific fields to study the UFO phenomenon. The 12 members of the original MJ-12 can be found by looking at the, well researched, Top Secret / MAJIC preliminary briefing document to President Eisenhower dated November 19, 1952 from MJ-2 Admiral Roscoe Hillenkoetter.

Dr. Vannevar Bush, Roscoe Hillenkoter, James Forrestal, Nathan Twining, Hoyt Vandenburg, Detlev Bronk, Jerome Hunsaker, Sidney Souers, Gordon Grey, Donald Menzel, Robert Montague and Lloyd Barkner were the first 12 members of MJ-12.

Richard Helms later acted as an advisor to MJ-12 and was a contributing editor to the "red book" which was a compendium on extraterrestrial activity which was supplied to the President of the U.S. every 5 years.

Further confirmation of MJ-12 is a Top Secret Memo from Wilbert Smith to the Canadian Department of Transportation on November 21, 1951 concerning a small group in the U.S. headed by Vannevar Bush secretly studying flying saucer technology. He stated that this UFO study group has a secrecy classification higher than the hydrogen bomb project.

Earlier, Wilbert Smith had made a request to use the facilities of Canada's Department of Transport to study UFOs. The project was formally approved on December 2, 1950 with the intention to collect data about UFOs and apply any recovered data to practical engineering and technology. The ultimate goal of the project was to apply any findings on the subject of geomagnetism to the possibility of exploiting Earth's magnetic field as a source of propulsion for vehicles.

Smith and his colleagues in government believed that UFOs, if real, might hold the key to this new source of power. A small-scale undertaking, the project called Project Magnet, used DOT facilities, with some assistance from personnel at the Defense Research Board (DRB) and the National Research Council.

In June 1952 Smith issued a preliminary report arguing that UFOs likely came from intelligent, extraterrestrial sources and almost certainly manipulated magnetism for flight. A 1953 report reiterated these conclusions. Also in April 1952 the Canadian government established Project Second Story, a parallel UFO research project, with Smith also involved. It consisted of a group of scientists and military officers who met periodically to consider the UFO question and to recommend government action. Smith reported to Second Story on some of Project Magnet's findings and conclusions.

Smith believed UFOs were linked to psychic phenomena and believed himself to be in contact with extraterrestrial beings who communicated to him through telepathy. Smith wrote a number of articles for *Topside*, the publication of the Ottawa New Sciences Club which he founded, outlining the philosophy of the "Space Brothers" with whom he claimed to be in contact. The articles were later collected and published posthumously in 1969 under the title *The Boys from Topside*

Some of the UFO visitors that crashed at Roswell were not extraterrestrials. According to Dan Burish, a biologist who worked at Area 51 Sector 4 and Arthur Neumann, a Physicist at Livermore Laboratories with a high level security clearance, the people that crashed in their craft at Roswell were from the future of our own planet. They had evolved into the Greys from the present human race because of pollution, radioactivity, living in underground cities and hybridization. They had come back to try to change the course of history and place the planet on a more favorable time line.

Their saucer crash was an accident caused by high-powered pulsed radar upsetting their navigation system – later the military realized this and adapted the radar as a weapon to shoot down other UFOs.

Their mission went disastrously wrong – not just because they crashed, but because they had a device with them which was their only means, as an orientation device in time and space, to get them home and back to their own time.

The device was a little box, which was subsequently utilized by military scientists in various experiments. When the box was acquired and investigated by the military, this became a catastrophe in itself. Much later, scientists at Los Alamos discovered that it was a time machine. A program running on a super computer was able to analyze the device and they discovered that a display on the device was displaying a scene just before the saucer crashed. After more computer analysis they were able to back engineer the device and develop a cronovisor which could see into the past and the future.

It made the timeline problem many times worse, because this both introduced the time-portal technology to us at the wrong time... and also told the military what lay ahead. Arthur Neumann, who works at Laurence Livermore laboratories as a physicist and has a high enough security clearance to work on MJ-12 projects, could not stress too strongly how totally calamitous for us all the Roswell incident was. It was a major setback, right at the start of the future humans' project to help fix the problem.

The live ET from the planet Serpo of the Zeta Reticulum star system was taken from the Corona crash site, established communications with us and provided us with a location of his home planet. The entity which was classified as EBE1 (extra biological entity #1) and their race Ebans, remained alive until 1952, when he died. But before his death, he provided us with a full explanation of the items found inside the two crafts. One item was a communication device. EBE1 was allowed to make contact with his planet.

President John F. Kennedy initiated Project Crystal Knight which was a continuation of the Project Plato ET diplomatic program with the added concept of a cultural exchange between our Planet and theirs. After the assassination of Kennedy, President Johnson continued Project Crystal Knight.

Later, a meeting date between MJ-12 and the beings of planet

Serpo, using the ET communication device, was set for April 1964 near Alamogordo New Mexico. The Ebans landed and retrieved the bodies of their dead comrades. Information was exchanged. Communication was in English. The Ebans had a translation device.

In 1965, we had an exchange program with the Ebans. We carefully selected 12 military personnel. They were trained, vetted and their records carefully removed from the military system. The 12 were skilled in various specialties.

Near the northern part of the Nevada Test Site, the aliens landed and the 12 Americans left. One Eban, EBE2, was left on Earth. The human team members were taken to an enormous mother ship which then traveled through wormholes or space tunnels which the Ebans knew about and arrived at Serpo nine months later.

The original plan was for our 12 people to stay 10 years on Serpo and then return to Earth. But something went wrong with their time calculation in the alien star system. The 12 remained until 1978, when they were returned to the same location in Nevada. Eight men returned. Two died on the alien's home planet. Two others decided to remain, according to the returnees. On their return, the project was renamed Project Serpo.

The returnees were isolated from 1978 until 1984 at various military installations. The Air Force Office of Special Investigation (AFOSI) was responsible for their security and safety. Of the eight that returned, all have died. The last survivor died in 2002. (More information at: http://www.serpo.org)

One interesting piece of technology given MJ-12 by these aliens came to be called the "Yellow Book". This was a viewing device somewhat like todays tablet computers that would display history. It could not only display the history of the Earth, but that of the whole Galaxy, but it would take many life times to view it all.

In July, 1952, there was a massive overflight of Washington D. C. by flying saucers. This went on for days. Thousands of eye witnesses observed them and stories and photographs appeared in the major newspapers.

Four months later, former General, Dwight D. Eisenhower was elected President of the United States. When Eisenhower replaced Truman as President he was briefed on UFOs, MJ-12, and the extra-terrestrial situation as described in the "Eisenhower Briefing Document."

Eisenhower ordered the CIA's Office of Scientific Intelligence (OSI) to determine if UFOs were interstellar vehicles. OSI convened the Robertson Panel which determined that UFOs could possibly represent a National Security threat. The panel recommended that the Air Force study UFOs under Project Blue Book.

Also, the panel recommended that a program of down-playing the UFO phenomenon to the public be started. This was because of their concerns that public belief in extra-terrestrial visitations would have a destabilizing influence on society.

In my previous book, *The Secret History of The New World Order*, I outlined the history of the creation of the CIA by Knights of Malta agents of the Vatican. The Vatican has considerable secret influence over the CIA and U.S. policy.

In any case, it was the Vatican that mainly wanted the UFO subject down played because it was feared that universal knowledge that extraterrestrials were visiting our planet, would undermine the Vatican's own mission, especially when the ET's had devices that could view the past and could prove that the present Christian religion was not what Christ had intended.

That is why the CIA's, OSI created, Robertson Panel made this decision. From that time on, the subject of UFOs has been ridiculed and downplayed in the controlled, mainstream news media and by the government. The Psychological Strategy Board (PSB) was created under the National Security Council to coordinate government-wide psychological warfare strategy." The PSB was also used to debunk UFOs to the U.S. public. On September 2, 1953, the PSB was replaced with the more powerful Operations Coordinating Board (OCB) which among other things, continued to debunk UFOs until the OCB was officially abolished by President Kennedy's Executive Order 10920 on February 18, 1961.

When Eisenhower replaced Truman as President he was briefed on UFOs and the extra-terrestrial situation as described in the "Eisenhower Briefing Document

At first, Eisenhower was worried about keeping the lid on MJ-12 and IPU secrets. On January 24, 1953, Eisenhower established the Advisory Committee on Government Organizations. He hired Nelson Rockefeller to be chairman of the committee. Rockefeller passed on to Eisenhower his recommendations on reorganizing the National Security Agency to protect the secrecy of MJ-12. MJ-12 should be headed by the Director of Central Intelligence.

Rockefeller was also appointed Special Assistant for Cold War Planning by Eisenhower and therefore also worked directly with the covert operations division at CIA headed by Frank Wisner.

Rockefeller ensured that MJ-12 now would have greater autonomy from the office of the president and the uncertainties of the political process.

The problem with MJ-12 working out of Wright Patterson was that it was a military base and therefore under the control of the U.S. President. In 1955, a remote section of the Atomic Energy Commission test site in Nevada was handed over to the CIA and called Area 51. It was an ideal test site for secret spy planes to be used against the Soviets.

Richard Bissell, working under Frank Wisner at CIA's covert operations was placed in charge of the acquisition of Area 51. Groom dry lake was an ideal place to build lengthy airstrips for testing their spy planes. Lockheed Aircraft Corporation was given access to Area 51 by the CIA to test their secret aircraft.

On the other side of Papoose Mountain, approximately 10 miles south-west of Groom Lake was Papoose dry lake. This was where MJ-12 would relocate in an area called S-4. S-4 was hidden inside of Papoose Mountain. There were multi levels of the underground facility. The first level housed the secret planes and flying saucers, which were in hangars that looked like part of the mountain from outside. Hangar doors could be opened and planes or saucers could fly right out of the mountain. The second level was the dining and meeting area. The third level down housed MJ-12 personnel. The fourth level down housed the aliens. The fifth level was a clean room laboratory where aliens and humans worked together on various classified technologies and genetics.

The CIA had total responsibility for the security at Area 51 and total control on who had access to the facilities there. The CIA's Directorate of Plans had the resources and personnel to relocate the extra-terrestrial technologies to S-4 and the CIA's Counter Intelligence Division made sure this was done without any leaks. Now, total secrecy was insured and even the U.S. president had no idea what was going on at Area 51, much less at S-4.

Later, Eisenhower would regret taking the office of the President out of the loop with MJ-12. He wanted to know the progress of the alien program. In 1958, a CIA agent and his boss were called into oval office. Eisenhower was accompanied by the Vice President, Richard Nixon. The president explained:

> "We called the people in from MJ-12 from Area51 and S-4, but they told us that the government had no jurisdiction over what they were doing…I want you and your boss to fly out there. I want you to give them a personal message…I want you to tell them, whoever is in charge, I want you to tell them that they have this coming week to get into Washington and report to me. And if they don't, I'm going to get the First Army from Colorado. We are going to go over and take the base over. I don't care what kind of classified material you got. We are going to rip this whole thing apart."

The CIA agent and his CIA boss went to Area 51 and relayed Eisenhower's message. Then, they were allowed to observe the flying saucers and Grey aliens at S-4. They interviewed the Grey alien. After returning to Washington D.C., The CIA agent and his boss reported what they had learned to the president, who was this time accompanied by J. Edgar Hoover, director of the FBI. The CIA agent observed that Eisenhower was visibly shaken from what he had learned. (7)

Secretive portions of the Military Industrial Complex were

working with the CIA and aliens on programs that the U.S. President had little or no control over. Future presidents would have even less ability to discover what was going on than Eisenhower had. So, effectively MJ-12, and corporations that it worked with, would become beyond the control of the constitutional government of the U.S. In Eisenhower's farewell speech, he warned about the dangers of the Military Industrial Complex.

According to William Cooper, of Naval Intelligence, in 1953, several large Alien spacecraft placed themselves in a high equatorial Earth orbit. Through electronics communications projects Sigma, communications were established between the U.S. Military and the alien craft. Project Plato was to establish diplomatic relations with this race and other races of space aliens.

However, another Alien group had set up a meeting between themselves and the U.S. Government first.

On February 20, 1954, President Eisenhower went to Muroc Air Field (Now Edwards AFB) accompanied by military Generals, reporter for the Hearst Newspaper Group, Franklin Allen, Catholic Bishop of Los Angeles, James McIntyre, Gerald Light of Borderlands Research, Edward Nourse of Brookings Institute and advisor to former President Truman and others for an unusual meeting.

All the men had to spend six hours in a security section being thoroughly interrogated about their lives, beliefs and other private matters of their lives before being admitted to the ET meeting area.

While at Muroc Field, five extraterrestrial space ships had landed, two cigar shaped ones and three saucer shaped ones. The extraterrestrials met with Eisenhower and his entourage. An ET translating device was used to communicate between the ETs and Humans. The ETs explained that they were from the Pleiadian Star system. They were human looking, tall, had blue eyes and long blond hair. Later, the military would dub them as "Nordics".

Electrical devices failed to work in the vicinity of the five alien craft present. So, the meeting was filmed with 4 spring wound cameras that had to be rewound every three minutes.

The Catholic Bishop, James McIntyre dominated the conversation. He wanted to discover the ET's religious beliefs. They informed him that every universe had a God. They had the true record of Jesus Christ. They stated that the first 4 Gospels of the New Testament were in error. Jesus survived his crucifixion and lived to an old age. They also stated that there was no Hell where souls would burn eternally. They explained that concept, like the resurrection, merely was an invention by the Church to give priests more power and influence.

The military generals were getting impatient with all this religious talk and wanted to discuss ET technology. So the ETs demonstrated their prowess by disappearing right in front of everyone's eyes and later, re-appearing. They asked some of the men present to lift up their flying saucer by hand. The humans were impressed how light the craft was as they easily lifted it up.

The military people wanted the Pleiadians to show them how their technology worked. The Pleiadians refused stating that the people of the earth were not spiritually evolved enough to handle the technology they already possessed. They also warned that the people of the earth were on a course that would eventually destroy their planet.

The Pleiadians also warned Eisenhower about the Aliens in the high equatorial orbit about the Earth, stating that their motives would be not for the benefit of mankind. They also asked that the U.S. stop all nuclear testing because of the danger to the Earths ecosystem and damaging effects on other dimensions.

They also asked Eisenhower to make their presence known to the public as soon as possible.

Eisenhower replied that humankind was not ready to handle this stupendous reality and needed time to be prepared for this kind of knowledge. So, the meeting was kept secret. However, Gerald Light, decided to inform his boss at Borderland Research and that is how this ET meeting was revealed to a few outside of government control. Gerald light would refer to the Pleiadians as "Etherians."

After the meeting, James McIntyre went to the Vatican and informed Pope Pious XII of the secret meeting with Aliens. Written transcripts and film records were stored in the Vatican library. Cardinal Spellman and Cardinal Cody, in the U.S. were also informed.

At the first Bilderberg Group meeting in May 1954, which Eisenhower attended, the item of extraterrestrial contact was discussed. Later, the Bilderberg Group discussed the extraterrestrial issue with the Council on Foreign Relations in secret meetings. Both of these groups are part of the New World Order's planning organizations.

By the mid-1950s these two groups agreed to give the Pleiadians an Island in French Polynesia as a base on Earth. They freely walk among the natives of the island.

This provided a way for extraterrestrials and Earth people to monitor each other and each other's cultures. And, when the NWO thinks the time is right, they will introduce the visitors from the Pleiades as ambassadors to the people of the Earth – at least that was what they said at the time. I am still waiting!

One of the first problems was that the Pleiadians offered to show Eisenhower, as they had with Roosevelt, how to create a paradise on planet Earth by offering to assist with a number of environmental, technological, political and socioeconomic problems but first he had to give up all atomic bombs and renounce war. Eisenhower feared this was a ploy to disarm the governments of the planet to make us easier to invade. So, he refused and no agreement was concluded at this meeting.

Later, in April 1954, Eisenhower would meet with the extraterrestrial group orbiting the Earth from the Betelgeuse star system in the Orion constellation, which the military would call the large nosed Greys. This group offered to exchange their technology in exchange for bases of their own on Earth. Also, they wanted to be able to do genetic testing on a few humans because they were a dying race that was having reproductive problems and they hoped to genetically correct their problem by gene splicing experiments with humans. The Zetas did not require the U.S. to give up their atomic weapons or give up war.

So, Eisenhower decided to do business with them with a secret treaty called the Greada treaty. This treaty is unlawful because it was not presented to the Senate and hence was not ratified by the Senate as required by law. Also, some of the terms of the treaty were unconstitutional.

The treaty stated that the aliens would not interfere in our affairs and we would not interfere in theirs. We would keep their presence on Earth a secret. They would furnish us with advanced technology and help us in our technological development. They would not make any treaty with any other Earth nation. They could abduct humans on a limited and periodic basis for the purpose of medical examinations and monitoring our development, with the stipulation that the humans would not be harmed, would be returned to their point of abduction, would have no memory of the event, and that the alien nation would furnish MJ-12 with a list of all human contacts and abductees on a regularly scheduled basis.

The Greys were given an underground base at Dulce, New Mexico on the Jicarilla Apache Indian reservation. Some Greys were later taken to Unit S-4 at Area 51 and an underground base at Indian Springs, Nevada.

The Dulce Base was located almost 2 miles beneath the Archuleta Mesa near Dulce. It was the first joint U.S. government – Alien biogenetics laboratory in the world and was highly classified under NSA and the CIA control. There were 7 levels to this base with the upper levels controlled by humans and the lower levels controlled by the Greys and Reptilians.

However after a period of time, NSA discovered things that alarmed them. First, there were a lot more abductions occurring than were being placed on the list and some of these weren't being returned. Second, they discovered that some of these female abductees were being held prisoner in level 7 of the Dulce Base and forced to have sex with the reptilians.

It turns out that a number of alien races consider Earth females to be very desirable. Not only can Reptilians have sex with female Humans, but they can also get them pregnant. The offspring will become hybrids, part Reptilian and part Human.

Also, there was some pretty horrible biological experimentation being done on humans at the Dulce base.

This intelligence got the men at NSA pretty riled up. NSA created two very secret organizations NSA Department X and Department Z. Department X, the intelligence arm, was tasked with keeping an eye on the Alien situation and Department Z, was the action arm.

When President Carter became inaugurated, he was briefed on the situation and was, like wise, very angry about the situation at the Dulce base. Department Z was tasked with solving the Dulce Base problem. Brigadier General Harry C. Aderholt organized, in October and September 1979, the group that would invade Dulce Base.

Captain Mark Richards was chosen to lead the attack team comprised of Air Force Special Operations Command, Navy Seals, Delta forces and Army rangers. Mark Richards had just been involved in a victorious space battle between some invading Reptilians and the Air Force Space Command in August of 1979 and was well qualified for this position.

The attack plan was to disable the primary generator for the base and free as many prisoners as possible. The operation was totally "black" not even the Joint Chiefs of Staff at the pentagon knew about it. The money for the operation came from Texas Businessman Ross Perot, CIA/DIA front man Edwin Wilson and a massive Black ops fund long hidden by Major E.L. Richards Jr., head of International Security. E.L. Richards Jr. was also Captain Mark Richards's father.

John V. Chambers who had been involved with Bechtel in the construction of the Dulce Base provided the attack team with a number of weak points in the base that could be attacked. Some of the reptilians had problems with certain human bacteria, like hay fever that could prove lethal to them. They required special filters on their levels to keep these bacteria out. So, one strategy was to disable these filter systems.

Further details of the Dulce War can be found at: http://www.anglefire.com/ut/branton/THE_DULCE_BATTLE.htm

One witness of this conflict, Phil Schneider claims that he was only one of three survivors of this conflict. Philip Schneider was an ex-government structural engineer who was involved in building underground military bases around the United States including the Dulce facility. He had a Level 3 Security Clearance, Rhyolite 38, which most have never heard of, but applies to geological operations underground.

Finally Schneider got fed up with the level of secrecy and resigned from his government job and started going on public speaking tours. He felt, as I do, that the American People have a right to know what their government is doing in secret. He became friends with Alfred Bielek, of Montauk Project fame, on these tours and the two traded information. Finally, Phil Schneider was murdered to silence him. Before his death, this is some of what Phil Schneider had to say:

"To give you an overview of basically what I am, I started off and went through engineering school. Half of my school was in that field, and I built up a reputation for being a geological engineer, as well as a structural engineer with both military and aerospace applications. I have helped build two main bases in the United States that have some significance as far as what is called the New World Order. The first base is the one at Dulce, New Mexico. I was involved in 1979 in a fire fight with alien humanoids, and I was one of the survivors. I'm probably the only talking survivor you will ever hear. Two other survivors are under close guard. I am the only one left that knows the detailed files of the entire operation. Sixty-six Secret Service agents, FBI, Black Berets and the like, died in that fight. I was there.

"Number one: part of what I am going to tell you is going to be very shocking. Part of what I am going to tell you is probably going to be very unbelievable, though, instead of putting your glasses on, I'm going to ask you to put your 'skepticals' on. But please, feel free to do your own homework. I know the Freedom of Information Act isn't much to go on, but it's the best we've got. The local law library is a good place to look for Congressional Records. So, if one continues to do their homework, then one can be standing vigilant in regard to their country.

"I love the country I am living in more than I love my life, but I would not be standing before you now, risking my life, if I did not believe it was so. The first part of this talk is going to concern deep underground military bases and the Black Budget. The Black Budget is a secretive budget that garners 25% of the gross national product of the United States. The Black Budget currently consumes $1.25 trillion per year. At *least* this amount is used in black programs, like those concerned with deep underground military bases. Presently, there are 129 deep underground military bases in the United States.

"They have been building these 129 bases day and night, unceasingly, since the early 1940s. Some of them were built even earlier than that. These bases are basically large cities underground connected by high-speed magneto-levity trains that have speeds up to Mach 2. Several books have been written about this activity. Al Bielek has my only copy of one of them. Richard Souder, a Ph.D. architect*, has risked his life by talking about this. He worked with a number of government agencies on deep underground military bases. In around where you live, in Idaho, there are 11 of them.

(*Richard Souder ~ not to be confused with Richard Sauder, Ph.D., an underground bases researcher and author of the book, 'Underground Bases and Tunnels: What is the Government Trying to Hide?')

"The average depth of these bases is over a mile, and they again are basically whole cities underground. They all are between 2.66 and 4.25 cubic miles in size. They have laser drilling machines that can drill a tunnel seven miles long in one day. The Black Projects sidestep the authority of Congress, which as we know is illegal. Right now, the New World Order is depending on these bases. If I had known at the time I was working on them that the NWO was involved, I would not have done it. I was lied to rather extensively."

Of further interest was Phil Schneider's father, Oscar Schneider, who worked as a Navy Medical officer and was involved with the Philadelphia Experiment according to Phil Schneider. After Oscar Schneider's death, a photograph was discovered in his basement showing the final briefing aboard the U.S.S. Eldridge on August 9, 1943 with Oscar Schneider among the other personnel. (8)

On May 1, 1975, another conflict between the Greys and humans took place at Area 51, unit S-4. The Greys were about to give a demonstration of how to extract energy from element 115. They were surrounded by many security guards with firearms. The Greys asked the guards to remove the ammunition from their guns as a safety precaution. The Greys thought that the antimatter reactions involved could trigger the guns to go off by themselves.

The guards refused to do that. The Greys insisted. One guard freaked out and shot a Grey. That was a big mistake, because the Greys have advanced psychic abilities which they used to kill all the guards present except one. That guard was left alive to tell the tale of what happened there.

That incident put a hold on technology transfers and cooperation between the Greys and MJ-12 for a period of time. According to Michael Wolf, the source of this account, MJ-12, renamed the Special Studies Group (SSG) by the NSC, had an advanced Alphacom Team that he was part of which was tasked with trying to diplomatically re- establish relations between MJ-12 and the Greys. Eventually, the Greys and MJ-12 came to an understanding that it was not a planned incident and everyone was reacting out of fear, rather than calm reason. They placed the incident behind them and restarted their cooperation.

Wolf states that Greys are masters of genetic engineering and that much of our present knowledge of genetic manipulation came from Grey technology. He worked on Project Sentinel with the Greys, which concerned cloning. Eventually, he successfully cloned a human being named "J-Type Omega". The idea was to create a super soldier that would be fearless and obey orders without question.

J-Type Omega was cloned inside a water tank from an embryo. Part of the DNA came from Wolf himself, making him a father of sorts. He was in the tank for one year before being "awakened." He emerged as a human of twenty years of age without any learning. He was intensely curious and was always asking questions.

When shown videos and films of Earth history, J-Type Omega cried when seeing war scenes with all their horrible cruelty. He had an extremely high IQ and told Dr. Wolf that he wanted to become a teacher.

Dr. Wolf programmed ethics into his creation after he realized that it had a soul. When ordered to kill a harmless dog, J-Type Omega refused. Dr. Wolf's superiors ordered him to terminate J-Type Omega, as a failure. With some help from a friend, Wolf was able to smuggle J-Type Omega outside Area 51 and set him free, telling him to never contact him again because it would be very dangerous for them both.

According to Dr. Wolf, a Star Gate was discovered inside a Giza pyramid. Scientists believe it to be some sort of lens which can create wormholes to any part of the universe. It hasn't been activated at the time because "We need to know a lot more about this galaxy and others before using it – where you want to go and how to prepare yourself. It's all about what you dial in. Coordinates have to be established. The problem is there are no traffic signs out in space." Dr. Wolf felt that this would be one of the last projects we will learn about through our association with the ETs.

Wolf noted that the Star Visitors are not comfortable with the world money and power brokers need for economic growth and increasing industrialization. They see these trends as destructive to the planetary ecosystem. Actually one of the Alphacom Team's missions is to discover some extraterrestrial technology could be used to help restore the planet's ecosystem to a more pristine state.

Another Alphacom Team mission is to determine the number and types of extraterrestrial visitors, their agendas and how to negotiate with them. According to Dr. Wolf, several confederations of ET civilizations are visiting us in a semi coordinated fashion. These would be the Alliance of the Altair Aquila system, The Zetas, The Federations of Worlds (from many star systems), and the United Races of Orion. The overall organization is called the Star Nations.

Dr. Wolf learned that in a high level briefing in the UK, the Vatican applied pressure to then U.S. President William Clinton to not make a planned announcement of the extraterrestrial visitations. Of particular concern to the Vatican was whether Clinton would make reference to how religion was created and why. The Vatican feels quite threatened by this whole subject. Dr. Wolf feels that the majority of humans are becoming more spiritual but that evil is also gaining in strength. This is one reason that he decided to come forward with what he knew.

Michael Wolf said that it took fifteen years of persuasion before his bosses finally allowed him to publish his book *Catchers of Heaven*. Their conditions were that the book had to be presented as a fiction and it had to have three different denials at the beginning about its authenticity. The book is basically about his work in these highly classified programs and gives a glimpse into the UFO cover up and ET reality.

Wolf had worked with Carl Sagan to decipher "The Monolith" first discovered in 1961 by both Yuri Gagarin and Alan Shephard, orbiting out in space in 1961 and recovered to Earth in 1972. The Monolith was constructed by ETs and emitted light and tone signals. If one were to close one's eyes and listen to the tones one would see a 3-d movie of the galaxy like you were in the film. Later, Wolf would see the same scenes taken from the Hubble telescope.

Carl Sagan could not talk about his work with extraterrestrials. His superiors at Cornell University would cut off the funding to his department if he did.

It is important to understand the level of compartmentalization of these secret projects. Often military personnel are used on covert projects without their commanding officers even knowing. They are sent TDY on another mission or training program, which is totally off the record. The CIA or NSA actually controls these covert missions.

Civilian contractors from DOD contracting companies are required to sign draconian non-disclosure contracts and reminded of laws that involve $10,000 fines and up to 10 years in prison for disclosure of classified information.

If anyone in these covert programs decided to become a "whistle blower" the usual procedure is to make all records of that person, birth certificates, educational records, employment history Etc. "disappear." Usually this means that person will suffer an end to their career. And sometimes whistle blowers end up "suicided" or just "disappear" themselves. So, becoming a whistle blower is not an easy decision for these people to make. Otherwise, there would probably be a lot more of them.

In one example of this secrecy, Vice Admiral Tom Wilson was J-2 head of intelligence for the Joint Chiefs of Staff at the Pentagon. Dr. Steven Greer of the *Disclosure Project* turned over some documents about classified extraterrestrial related projects which included a list of code names and project names over to Wilson. When Wilson checked to determine if the projects really existed, he was denied access.

Greer said that when Wilson identified the group, he told the contact person, "I want to know about this project." And, he was told, "Sir, you don't have a need to know. We can't tell you." The contact persons were not DOD. They were attorneys for the defense contractors! (9)

So, here we see an example of what President Eisenhower was warning the American people about in his farewell speech. The Military Industrial Complex, or parts of it, that neither the President nor the Pentagon have knowledge of or control over!

Besides meeting with government leaders particularly U.S. leaders after the war, the E.T.s met with regular people who became known as "contactees". One famous Swiss contactee is "Billy" Eduard Albert Meier. Billy Meier has some of the best Flying Saucer photographs in the world. Photographic experts from Japan have analyzed his photos and found them to be authentic. And, many of these photos were taken before home computers and Photoshop.

He presents these photographs as evidence to support claims that he is in contact with extraterrestrials. In addition, he has also presented other controversial material during the 1970s such as metal samples, sound recordings and film-footage. Meier reports regular contacts with extraterrestrials he calls the Plejaren. Meier claims that the Plejaren look similar in appearance to humans from Northern Europe (as do the Pleiadians), and states that the Plejaren home world is called Erra. It is located in a dimension which is a fraction of a second shifted from our own dimension, about 80 light years beyond the Pleiades, an open star cluster.

This information is quite interesting because if true, it could mean that time holds the possibility of shifting dimensions, leading to the possibility of traveling over a hundred light years in a fraction of a second. Torsion waves and scalar EM waves, as Thomas Bearden and Nikolay Kozyrev have stated, also have this ability.

In any case, Meier claims to have befriended some of these Plejaren, in particular one woman named Semjase, the granddaughter of Sfath. Meier says that he has also had many contacts with another Plejaren man called Ptaah, starting in 1975 and continuing right up to the present day. Semjase has invited Billy Meier onboard her "Beam Ship" and taken him on a number of trips in outer space and through time.

He has claimed that he has also visited other worlds and galaxies along with another universe with these Plejarens. Meier claims that he was instructed to transcribe his conversations with the various extraterrestrials, most of which have been published in the German language. These books are referred to as the *Contact Notes* (or *Contact Reports*). Currently, there are twenty six published volumes of the *Contact Reports* (titled *Plejadisch-Plejarische Kontaktberichte*).

Some of the *Contact Reports* were translated into English, extensively edited and expurgated, and published in the out-of-print four-volume set *Message from the Pleiades: The Contact Notes of Eduard Billy Meier* by Meier case investigator, Wendell C. Stevens. Wendell C. Stevens, a retired Air Force Intelligence officer, has done considerable UFO research. He has written many books on the UFO subject based on his researches with many contactees. Many of these books are available at Amazon.com.

There are also many contact reports translated into English (unedited) by Benjamin Stevens. His website is here: http://billy-meiertranslations.com

Billy Meier's latest writing, *The Goblet of Truth* an interplanetary teaching on universal spiritual wisdom is available in both German and English for free at: http://www.figu.org/ch/files/downloads/buecher/goblet-of-the-truth.pdf

Other contactees included George Adamski, Daniel Fry, Truman Berthurum and George Van Tassle in the U.S., who claimed they were communicating with the beings in the flying saucers. Many contactees have written books, some with photographs, about their experiences and communications with these alien beings.

Van Tassle built the Integratron at Giant Rock, California. The Integratron was supposed to be a healing building based on information from the ETs. VanTassle died before completing the Integratron. The building was eventually sold to a New Age Group that had no idea how to make it work.

A notable U.S. contactee was Howard Menger who, in his book, *From Outer Space to You,* describes his contacts with ETs which took place over a long period of time starting in the 1940s. His pictures of the flying saucers were similar to the Adamski pictures and the appearance of the ETs both contacted were human looking, much like the Pleidians that contacted Eisenhower. However, they stated that they were from Mars and Venus. And, they had a program of contacting select people directly, in addition to government leaders. Those selected were more spiritually advanced and did not suffer from false ego problems.

Howard Menger was told by his ET friends that they had an important mission for him and that after 1957, he was supposed to go public with his meetings with the extra-terrestrials. He befriended GeorgeVan Tassle and both appeared on the radio show talk show *The Party Line* by Long John Nebel on Station

WOR in New York, during and after 1957, bringing information to millions of the public that we were being visited by beings not of this world.

At one point his ET friends took, Menger on a trip to Venus. It was a quick trip that only took minutes to arrive there from Earth.. As they did a fly-by of the surface of the Planet, Menger looked out a viewing port and saw scenes of their civilization there. There were animals that he didn't recognize, buildings and trees there and the place looked quite beautiful to Howard. They didn't get out of the craft, however, and soon returned to Earth.

He also took a trip to the Moon with some other contactees, some who Howard Menger knew, to a base on the Moon. There they landed and got to explore the Moon for a number of days before returning to Earth. There were other contactees from other countries at the domed Moon base. So there was an ET program to educate the Earth's people of their presence, while the governments were trying to keep it all secret.

Later, Howard Menger, as described in his book, *The High Bridge Incident: The Story behind the Story,* was able to build a radio controlled model flying saucer based on information from the ETs. He called his saucer an Electro Craft. One day, while flying the Electro Craft near High Bridge, New Jersey, it flew out of range of his radio control transmitter and was lost.

A few weeks later, the FBI knocked on his door. They had the remains of his crashed Electro Craft, which had crashed in Ohio, and asked him if this belonged to him. The FBI had tracked the part numbers of components of the craft back to venders that had sold the parts. Some of the vendors could tell the FBI who the parts were sold to and the FBI was to eventually determine who had built the craft.

After that, Howard Menger was asked to go to certain government locations and reveal to scientist there the working principles of his Electro Craft. Then, Howard Menger was told to cease from building any more Electro Craft because of National Security concerns. Being a patriotic American, Menger agreed.

Also there are contactees from other nations besides Billy Meier. In his book *Mass Contact*, Stefano Breccia not only outlines the Ummo affair, dealing with beings from the planet Ummo, but also outlines long lasting contacts between himself and others in his group with another race of extraterrestrials that lasted over 40 years in Italy. The contactees were taken to large extraterrestrial underground bases in Italy and witnessed flying saucers and teleportation on a number of occasions.

The National Aeronautics and Space Administration (NASA) has played a large role in the cover up of the extraterrestrial presence. As previously stated, NASA is a smoke screen to hide antigravity technology by making the public think that rockets are the only way to get into space.

NASA also has gone to great lengths to hide evidence of artifacts of alien civilizations discovered on the Moon and Mars.

Although it was publically stated that NASA was a purely civilian scientific agency, investigators have pointed out in the fine print of the NASA charter that it also works in conjunction with the Department of Defense and some NASA findings can be classified information for reasons of National Security.

After the findings of the Robertson Panel and studies by the Brookings Institute, it was decided that public knowledge of the extraterrestrial presence would have a destabilizing effect on society. Hence it was decided that secrecy of the extraterrestrial presence was a matter of National Security. Even ancient artifacts evidencing past ET civilizations on Earth, the Moon or other planets was considered a matter of National Security. These are the reason for NASAs cover up of extraterrestrial Civilizations evidence on the Moon and other planets.

One interesting NASA researcher is Richard C. Hoagland, who coauthored with Mike Bara the book *Dark Mission The Secret History of NASA*. In the fascinating book, photographic evidence for ancient extraterrestrial civilizations on both Mars and the Moon are presented.

Hoagland was studying the Viking photos of the Cydonia region on Mars where the famous Face on Mars is located. He noticed a number of tetrahedral pyramids and a pentagonal pyramid in the surrounding region. He drew lines connecting the various artifacts and noticed a reoccurrence of many mathematical ratios and angular measurements as had been discovered among the Egyptian pyramids at Giza. One reoccurring angle was 19.47 degrees.

While pondering the significance of this particular angle and the tetrahedron pyramids, he realized if tetrahedrons were enclosed in spheres such that if one apex of the tetrahedron is at the pole of the sphere the other three apexes will be at the 19.47 degree latitude of the sphere.

This 19.47 latitude, either north or south, seems to be important in planetary physics as this is the latitude of the large shield volcanoes on Earth at the island of Hawaii and Mars at Olympus Mons. The Red Spot on Jupiter is also at this latitude and sunspots are clustered at this latitude on the Sun.

Hoagland then became acquainted with torsion field theory as developed by Nikolai Kozyrev. He combined the two concepts and developed a hyper dimensional physics. Hoagland's hyper dimensional physics had to include the total angular momentum of a planet, its rotational and orbital angular momentum. Using this theory, he could predict the anomalous infrared (IR) heat emission from the outer planets before it was measured by fly by satellites.

Energy from higher dimensional space is brought into our three dimensional space through angular momentum. And this additional energy was being radiated away from the outer planets in larger amounts that being taken in by incident solar energy. That was the source of the anomalous IR energy that was successfully predicted by Hoagland's hyper dimensional physics. This hyper dimensional theory could bring breakthroughs in other areas of science and should be thoroughly explored and tested.

Another interesting fact that was revealed in *Dark Mission The Secret History of NASA* was the number of 33 degree Scottish Rite Freemasons of the Southern Jurisdiction involved in NASA. These illuminati members are ultimately controlled (knowingly or unknowingly) by the Jesuit General at the Vatican.

Also, we know that a number of NASAs directors and employees were former members of the Nazi SS, like Werner Von Braun, Humbertus Strughold and Kurt Debus. Himmler, himself a Jesuit, created the Nazi SS to be like the Society of Jesus or the Jesuits and was also controlled by the Vatican. So, the Vatican's desire to keep the extraterrestrial presence secret is definitely a part of NASA policy.

Aldrin was a Mason who took a Masonic apron and a 33 degree Scottish Rite Freemasons of the Southern Jurisdiction flag to the Moon and posed with it for a photograph on one of his EVAs. This photograph is on the front cover of *Dark Mission The Secret History of NASA*.

Another interesting whistle blower was ex MJ-12 (now called Special Studies Group) member, Dr. Michael Wolf. He stated that he worked daily with Zetas at a secret underground facility. He also claims that the U.S. administration entered into a secret treaty with the extraterrestrials from the Star System Zeta Reticuli in the 1950s. This secret treaty is different than the Greada treaty with the Grays from Betelgeuse. Of course, this treaty is also un-lawful since Congress was never informed of it and the Senate never ratified it.

Dr. Michael Wolf has written some fairly informative pieces on his own work with extraterrestrials including a book titled *The Catchers of Heaven*. Wolf says that there are many ET groups visiting the Earth like the Zeta Reticulum's, ETs from the Altair Aquila system, The United Races from Orion and the Federation of Worlds (or Star Nations), from many star systems.

The information provided by Dr. Wolf may be part of a gradual release of information to the public to acclimate them to the extraterrestrial presence. Other sources of this acclimation program likely include Phillip Corso, author of *The Day after Roswell*, Air Force Colonel (Retired) Steve Wilson and Air Force Colonial (Retired) Donald Ware.

The CBS show *Unsealed Alien Files* is also a means of gradual public acclimatization to the alien presence, although the program uses a fear based presentation that barely reveals the tip of the Alien presence ice burg. If the ETs were bent on conquering Earth, they would have done it long ago, before we developed the weapons we have now.

Wolf also has stated that many powerful institutions including the Vatican, oil companies and pharmaceutical companies are opposed to this release of information. Free energy and ET healing technology would definitely cut into the cash flow of the oil and pharmaceutical companies. The Vatican fears losing followers, since the ETs have definite proof that the present Christian religion is not what Christ intended.

Wolf reveals that there is a cabal of plotters top-heavy with military and headed by an Under-Secretary of The Navy, that secretly undermines the goal of peaceful dealings with the ETs. They sometimes use weapons, developed under the Strategic Defense Initiative, to shoot down ET saucers.

So, the extraterrestrials are here, but a cabal wishes to hide that fact from the world's people. And, if you have read my *Secret History of the New World Order*, you know that, through their agents, the Vatican created and controls the CIA and all of the intelligence agencies of the World. And, these intelligence agencies are trying to keep the lid on the extraterrestrial presence.

6

Exopolitics

Exopolitics is a direct logical extension of conventional politics to the interplanetary theatre. Alfred Webre J.D., who first formally introduced Exopolitics as a discipline of study, defined it as the study of law, governance and politics in the Universe.

The need for exopolitics arises naturally. Once one accepts the existence of other sentient beings besides ourselves in our immediate neighborhood (which for the purpose of this discussion can be defined as the solar system and nearby stars), it becomes necessary to look at our neighbors politically. Like any sentient beings, they naturally have to have their own needs, desires, interests and agendas.

By analogy with political parties on Earth, which are aggregations of individuals who share a certain common political agenda, i.e., a certain set of policies they mutually seek to bring about, we can introduce the notion of an exopolitical party, which we shall define as any grouping of members of the interplanetary community with a specific exopolitical agenda, i.e., a specific set of policies toward other members of the interplanetary community.

The main job of exopolitics research then consists of identifying the existing members of our immediate neighborhood and classifying them by their exopolitical agendas, thus establishing a picture of the existing exopolitical parties in our immediate neighborhood.

As we have seen, Exopolitics has been secretly conducted by the U.S. and other Earth based governments. Much of this secrecy is now being breached by whistleblowers and diligent UFO researchers.

Also, there seems to be persons having high level security clearances that believe the public has a right to know what is going on. These people have taken upon themselves to allow a gradual release of information to a select portion of the public that they deem able to handle the information without panic. That portion of the public is represented by those that attend UFO conferences and read UFO contactee information Etc. and have been mentally acclimatized to the high probability of the extraterrestrial presence on Earth.

In other words, this segment of the public would be the kind of people that might read this book. Other segments of the public could be too involved with their own personal daily affairs to really care one way or the other or give it much thought.

And then, there are the types that would rather be in denial of the whole subject and keep their heads buried in the sand because it causes too much mental and/or emotional discomfort for them to deal with it.

I have personally known a lady that was abducted by the Greys on multiple occasions in her life, starting at age 14. The abductions occurred at night while she was asleep in her bed. A bright blue beam of light would shine right through the ceiling of her bedroom that would paralyze her and lift her right out of her bed. Then she would be taken right through the ceiling and roof of her apartment. When outside, she could look up and see a flying saucer overhead. Then, she would be taken through an opening in the bottom of the saucer into the craft and placed on a medical examination table.

She was later impregnated by the Greys at a time when she had no sexual relations with anyone. Her doctor verified that she was pregnant. Three months later, she was no longer pregnant. Her doctor verified that she no longer was carrying a fetus but saw no indication of an abortion. He was mystified. A year later, she was abducted again and while onboard the saucer, she was given a hybrid, part human, part grey, baby to hold. The baby looked ugly to her. She was informed that the baby was hers and that it needed human affection.

The whole affair traumatized her considerably and while she confided her story to me only after I promised not to reveal her name, she really tried to avoid the whole subject and would like to pretend that the whole affair never happened and aliens didn't exist.

Bret Oldham was abducted throughout his life starting at a young age. In his book, *Children of the Greys,* he gives detailed descriptions of his abductions. During some of these abductions, he witnessed other human abductees and one that was a personal friend of his. His wife Gina became pregnant even though she was taking birth control. Later, both he and Gina were abducted together and he was forced to watch as the fetus inside Gina was surgically removed from her by the Greys.

A few years later, Bret Oldham was again abducted and shown

a rather cute, half human half Grey young child. He was informed that it was his daughter. At first he didn't believe it. But, soon he started emotionally bonding with her. The visit didn't last long however and he never did see her again.

A similar scenario has happened to many others around the Earth. These people, for the most part, have been traumatized by these abduction experiences and would like to pretend that aliens didn't exist at all so they could lead a normal life.

Another factor in these abductions is *Millabs,* the joint secret military and Grey laboratories where some of these abductions take place. Many of the abductees have witnessed humans, usually in black uniforms, and greys working together during their abductions.

One primary purpose of these abductions is to create a hybrid race of part Grey part Human beings. This is because the Greys are losing the ability to sexually reproduce and also are lacking in human emotions.

These ongoing abductions are a part of the secret Greada Treaty between U.S. President Eisenhower and the Greys from Betelgeuse star system in the Orion constellation and therefore are part of exopolitics.

Some very well researched, well documented and quite thoughtful books on the subject of exopolitics have been written. *Exopolitics: Politics, Government, and Law in the Universe,* a book by Alfred Webre, cover many of the issues involved. Alfred Webre was the first to introduce the concept of exopolitics to the public of this planet.

Another book written by Alfred Webre is *The Dimensional*

Ecology of the Omniverse. The Omniverse includes the material universe, higher dimensional universes and the spiritual universe. The well documented evidence for this subject matter comes from thousands of hypnotic regression therapy sessions done by qualified practitioners around the world. Exopolitics also exists between different parts of the omniverse, as Webre points out in this book. An interesting website that contains some of Webre's subject matter is here: http://exopolitics.blogs.com/

Exposing U.S. Government Policies on Extraterrestrial Life by Michael Salla is a well-researched and documented work on U.S. Government secret relations with Extraterrestrials.

Another book by the same author, which I highly recommend reading if you want to learn more about the various beings and civilizations visiting earth and their interactions with our civilization, is *Galactic Diplomacy: Getting To Yes With ET*. In this book, an explanation for the devious conduct of the Military Industrial Extraterrestrial Complex (MIEC) is offered by visitors from Procyon, a binary star system 11.4 light years distant.

UFO researcher George Andrews interviewed a contactee who revealed this information from a human visitor from Procyon named Khyla. The information from Khyla had a high degree of correlation with information Andrews had obtained from other sources.

According to Khyla, the Procyon civilization flourished until Greys that populated Riegl attempted to subvert Procyon. Khyla described the process used by the Greys to subvert their civilization:

"The Greys began to visit us, first a few as ambassadors, then as specialists in various domains where their expertise could be useful to us, as participants in different programs that involved mutual collaboration, and finally as tourists. What had begun as a trickle became a flood, as they came in ever increasing numbers, slowly but surely infiltrating our society at all levels, penetrating even the most secret of our elite power groups… Just as on your planet they began by unobtrusively gaining control over key members of the CIA and KGB through techniques unknown to them, such as telepathic hypnosis that manipulates the reptilian levels of the brain, so on Procyon through the same techniques…they established a kind of telepathic hypnotic control over our leaders. Over our leaders and over almost all of us, because it was as if we were under a spell that was leading us to our doom, as if we were being programmed by a type of ritual black magic that we did not realize existed." (10)

A similar program of subversion seems to occurring on our planet. Some whistleblowers have claimed that an underground Reptilian base is under the Vatican and that their leaders are mind controlled. The Vatican leaders have seemed to be mind controlled since Christianity was made the legal religion of Rome, as their long bloody and cruel history has demonstrated.

The Greys obtained a foot hold before World War II when they contacted Hitler and Roosevelt, who entered into secret treaties with the Greys in return for technology. This occurred after they were previously contacted by Plaidians who tried to persuade them both to pursue a higher more peaceful spiritual path that would lead to paradise on Earth. History shows that both leaders chose the wrong path and sided with the wrong ETs that lead to hell on Earth with World War II.

After the War, Eisenhower also chose the wrong path and renewed the treaty with the Greys. Subversion of the secretive CIA and KGB heads by telepathic hypnosis led to more Hell on Earth as any historical study of these agencies will attest. Presently, the United State, after being on a nation destroying binge in the 21st century, is considered the greatest threat to world peace by the majority of the people of this planet. So obviously, some U.S. Leaders, Corporate CEOs and military leaders also have become mind controlled by forces that wish to destroy our planet or perhaps take it over.

According to contactee, Alex Collier, the Procyons have finally liberated their world from the influence of the Greys from Reigle by developing multidimensional consciousness and using mind imagery to protect one's self from extraterrestrial mind control. Perhaps we people of the Earth could also learn how to develop multidimensional consciousness and mind imaginary to do the same.

Extraterrestrial subversion can be countered by exposing global secrecy of the extraterrestrial presence and creating greater awareness of the problem. Leaders that violate national and international law, as our leaders seem so inclined to do, should be prosecuted. Law should apply equally to all members of society whether that person is a janitor, billionaire or president of a nation. No one should be above the law. If a person considers themselves above the law, commits a crime and gets away with it, it creates a greater incentive for those who follow to also do likewise.

According to the United Nation's Charter of which the United States is a member, a nation that conducts a military attack on a sovereign nation commits an international crime. How many times has the United States committed this sort of crime since the creation of the United Nations? The problem is that there is no United Nations method to punish the guilty unless they are members of a militarily defeated nation. So, might still makes right on Earth until a better way is found and U.S. war criminals, and those of other nations, continue to go unpunished.

Obviously the U.S. needs an *independent* of the White House Justice Department that can effectively prosecute high level crime. Congress has failed its job in that area. If all the Iran Contra criminals, including George H.W. Bush, Admiral Poindexter and Bill Clinton, were properly prosecuted and imprisoned for life, there probably would never had been a 9/11 attack, since some of the 9/11 players with powerful patrons, like the Vatican and their Masonic secret societies, knew that they would be protected, as were many of the Iran Contra criminals, and essentially considered themselves above the law.

The CIA is totally outside national and international law by being involved in political assassination, instigating insurrection, regime change and illegal narcotics trafficking and needs to be disbanded, as John F. Kennedy planned to do before he was "taken out." Enough of this hiding criminal activity under the cloak of National Security! Other intelligence agencies engaged in illegal activities, like NSA, DIA, Naval Intelligence AFIOC and DISC should also be purged and required to act within lawful constitutional limits. If not, the U.S. will always be considered a criminal nation by those of us, around the world, that actually believe in justice and law and order.

Michael Salla believes that the veil of secrecy needs to be lifted from exopolitics and that schools should offer courses in other planetary civilizations, and extra planetary diplomacy. I totally agree.

However we also need to clean up regular politics right here in Earth; if not why should other ethical ET civilizations want to do business with us? Birds of a feather flock together. And, it is a mistake to do business unethical ET civilizations, as history has already clearly demonstrated.

Other books by Michael Salla are *Exopolitics: Political Implications Of The Extraterrestrial Presence* and *Kennedy's Last Stand: Eisenhower, UFOs, MJ-12 & JFK's Assassination* Also, Michael Salla has a website on some of these subjects here: http://exopolitics.org/

Paradigm Research Group also is quite interested in exopolitics. This site is the United States component of the Exopolitics World Network - a project of Paradigm Research Group: http://www.paradigmresearchgroup.org/exopoliticsunitedstates.htm

One of the big issues of Exopolitics is a treaty to prevent the weaponization of space. The Star Nations and many Earth governments are working towards this goal. Both China and Russia are working towards a treaty. To learn more check out http://www.peaceinspace.com/the-treaty

7
Modern Antigravity Development

Townsend Brown left Swazey Observatory in 1930 and went to work for The Naval Research Laboratory in Washington D.C., as a specialist in radiation, field physics and spectroscopy. From 1931 to 1933, the Navy placed him in charge of a project to investigate unusual electric effects in fluids and high K dielectrics.

These researches showed that massive high K dielectrics had the best electrogravitic effects. These electrogravitic forces varied with planetary positions. He also discovered that certain of these dielectrics underwent resistive changes which changed with planetary positions. He made an instrument, with no moving parts, that measured the changing gravitational anomalies by passing current through these dielectrics to a galvanometer.

His observations were both at the Naval Laboratory in Washington D.C. and at Sea on the Submarine S-48 in the West Indies, as part of the Navy-Princeton International Gravity Expedition. These researches resulted in a, still classified, document titled, "Anomalous Behavior of Massive High-K Dielectrics."

Also, in 1933, Brown was given leave to join, as a physicist, another geophysical expedition to the Caribbean sponsored by the Smithsonian Institute and financed by Eldrige Johnson, cofounder of RCA, onboard Johnson's yacht, the *Caroline*. The stated purpose of this mission was to map underwater rifts.

On this cruise Brown became acquainted with Eldrige Johnson, his wealthy business partner, Leon Douglass and British spy master, William S. Stephenson. All 4 men became members of Stephenson's secret international group dubbed the "Caroline Group" which would play a large role in Brown's future.

After the job at Naval Research, Brown held a variety of jobs. In 1938, he served as an assistant engineer on the maiden voyage of the USS *Nashville*. On its return trip from Europe, the *Nashville* carried a cargo of $50 million in gold being transferred from the Bank of England to Chase Manhattan Bank in New York. While Brown was away on this voyage, Eldrige Johnson was overseeing the construction of an electrogravitics research laboratory at the University of Pennsylvania. Part of the money transferred by the *Nashville* was used to fund this laboratory.

After his Voyage on the *Nashville*, Brown went to work at the University of Pennsylvania at the laboratory constructed for him. About a year later, in 1939, Brown left The University of Pennsylvania and went to work as a materials process engineer for Glen Martin Company in Baltimore. Glen Martin would later become Lockheed Martin Aerospace Corporation.

In 1940, the Navy called him to head a "mine sweeping research and development project". After the attack on Pearl Harbor and the U.S. entry into World War II, Brown was assigned to the Naval Operating Base at Norfolk, Virginia. Then, he was sent to the Philadelphia, Navy yard to outfit new ships about the same time the infamous Philadelphia Experiment was taking place. When later asked about his involvement in that project his only response was "I am not permitted to talk about that part of my work."

Much effort was made to hide Brown's work record during that period of history with several different versions surfacing. It is claimed that Brown retired from the Navy in December of 1943 after suffering a nervous breakdown. Could his nervous breakdown have been caused by the disastrous results of the Philadelphia Experiment of August 12, 1943 where some of the crew members of the Eldrige were embedded into the steel of the destroyer escort after its reappearance?

In any case, in June of 1944, Thomas Townsend Brown went to work at the Advanced Projects Unit of Lockheed Vega Aircraft in Burbank, California, which was the forerunner of Lockheed's Skunk Works. From this, one can be sure that some of the Lockheed engineers knew all about the Biefield-Brown effect and Brown's researches in electrogravity.

After the War, Brown and his family set up the Townsend Brown Foundation in Los Angeles, California. In his laboratory there, Brown continued to improve his electrokinetic disks. In 1950, Brown was hired as a consultant by Admiral Arthur Radford, Commander in Chief of the U.S. Pacific Fleet at Pearl Harbor to demonstrate his electrokinetic disks.

Nothing seemed to result from the Pearl Harbor demonstration. In 1952, Thomas Townsend Brown submitted a proposal to the Navy, which they called Project Winterhaven, in which Brown urged the Navy to develop, in a secret program like the Manhattan Project, a flying saucer utilizing flame jet high voltage generators to propel the saucer up to 3 times the speed of sound. The technological principles were probably along the lines outlined in Brown's U.S. patent number 3,022,430.

Another part of the proposal was to develop massive high-K dielectrics to be used as super-efficient ship propulsion systems.

Another subject in the Project Winterhaven proposal was gravitational wave communication. Brown had developed a way to transmit gravitational waves by discharging high voltage capacitors through a spark gap. One side of the capacitor was grounded and the other side connected to a spherical antenna. The receiver was similar to his gravitational anomaly detectors.

In one demonstration of his gravitational wave communication device, in 1952 in Los Angles, he was able to transmit a signal over 35 feet to a receiver located inside a grounded metallic enclosure. Electromagnetic waves would not have penetrated this Faraday Cage. So Brown was convinced that they were gravitational in nature. His transmitter was highly similar to ones used in Tesla's experiments with spherical antennas. Could it be possible that Tesla's longitudinal waves were also gravitational waves?

These gravitational waves could be transmitted easily through conductive media like sea water and could be used to communicate with undersea submarines and underground bases.

On March 21 1952, while Brown was about to give a demonstration of his electrokinetic disks to some colleagues, he was unexpectedly visited by Air Force Major General, Vic Bertrandias who demanded to be included in the demonstration.

Vic Bertrandias was a former vice-President of Douglas Aircraft and a close friend of General Albert Boyd, director of Air Force Systems Command at Wright Air Development Center. It must be remembered that Wright Field was where some of the Nazis that worked on flying saucers were brought under Operation

Paperclip after the War. Some of these German scientists already knew all about Brown's work with antigravity and had already engineered it into operational craft. So, Albert Boyd probably wasn't that impressed with Vic Bertrandias' report of Brown's demonstration.

After Brown's demonstration to Vic Bertrandias, he became worried that the Air Force might try to classify his work. So, two weeks later, he arraigned for a press conference in Los Angeles. The reporters from the Los Angeles Times were invited to see a demonstration of his electrokinetic disks and later ran a story on the subject. So, in 1952, the public was informed of the very real existence of antigravity technology.

From 1952 until 1958, the press would run an occasional story on antigravity research being done by various air craft and electronics corporations. An informative story was run by the *New York Herald Times* on Sunday, November 20, 1955 by Ansel E. Talbert, Military and Aviation Editor:

> "This is a first of a series on new pure and applied research into the mysteries of gravity and efforts to devise ways to counter act it.
>
> The initial steps of an almost incredible program to solve the secrets of gravity and universal gravitation are being taken today in many of America's top scientific laboratories and research centers.
>
> A number of major, long-established companies in the United States aircraft and electronics industries also are involved in gravity research. Scientists, in general, bracket gravity with life itself as the greatest unsolved mystery in the Universe. But, there are an increasing numbers

who feel that there must be a physical mechanism for its propagation which can be discovered and controlled.

Should this mystery be solved it would bring about a greater revolution in power, transportation and many other fields than even the discovery of atomic power. The influence of such a discovery would be of tremendous import in the field of aircraft design – where the problem of fighting gravity's effects has always been basic.

One almost fantastic possibility is that if gravity can be understood scientifically and negated in some relative inexpensive manner, it will be possible to build aircraft, earth satellites, and even space ships that will move swiftly into outer space, without strain, beyond the pull of Earth's gravity field. They would not have to wrench themselves away through the brute force of powerful rockets and through expenditure of expensive chemical fuels.

Centers where pure research on gravity is now in progress in some form include the Institute for Advanced Studies at Princeton, N.J. and also at Princeton University: the University of Indiana's School of Advanced Mathematical Studies and Purdue University Research Foundation.

A scientific group from Massachusetts Institute of Technology, which encourages original research in pure and applied science, recently attended a seminar at the Roger Babson Gravity Research Institute of New Boston, N.H., at which Clarence Birdseye, inventor and industrialist also was present. Mr. Birdseye gave the world its first packaged quick frozen foods and laid the foundation for today's frozen food industry; more recently he has become interested in gravitational studies.

A proposal to establish at the University of North Carolina at Chapel Hill, N.C., an 'Institute of Pure Physics' primarily to carry on theoretical research on gravity was approved earlier this month by the Universities board of trustees. This had the approval of Dr. Gordon Grey who has since retired as president of the University; Dr. Grey has been Secretary of the Army, Assistant Secretary of Defense, and special assistant to the President of the United States.

Funds to make the institute possible were collected by Agnew H. Bahnson Jr., an industrialist for Winston Salem, N.C.. The new University of North Carolina administration is now deciding on the institute's scope and personnel. The directorship has been offered to Dr. Boyce DeWhitt of the Radiation Laboratory at the University of California at Berkeley, who is the author for a Roger Babson prize winning scientific study entitled "New Directions for Research in the Theory of Gravity."

The same type of scientific disagreement which occurred in connection with the first proposal to build the hydrogen bomb and an artificial earth satellite – now under construction – is in progress over anti-gravity research. Many scientists of repute are sure that gravity can be overcome in comparatively few years if sufficient resources are put behind the project. Others believe it may take a quarter of a century or more.

Some pure physicists, while backing the general program to try to discover how gravity is propagated, refuse to make predictions of any kind.

> Aircraft industry firms now participating or actively interested in gravity include the Glenn L. Martin Co. Baltimore, builders of the nation's first giant jet-powered flying boat; Conair of San Diego, designers and builders of the giant B-36 intercontinental bomber and the world's first successful vertical take-off fighter; Bell Aircraft of Buffalo, builders of the first piloted airplane to fly faster than sound and a current jet vertical take-off and landing airplane and Sikorski division of United Aircraft, pioneer helicopter builders.
>
> Lear Inc. of Santa Monica, one of the world's largest builders of automatic pilots for airplanes; Clark Electronics of Palm Springs, California, a pioneer in its field, and the Sperry Gyroscope Division of Sperry-Rand Corp. of Great neck L.I., which is doing important work on guided missiles and earth satellites, also have scientists investigating the gravity problem..."

Agnew H. Bahnson Jr., mentioned in the above article, invited Brown to his laboratory in North Carolina In November 1957 to do more research. Together with Dr. Frank King, they started experimenting with a combination of high voltage D.C. and alternating current fields. In one embodiment, a parabolic contoured disk was used as the upper electrode and a copper screen of the same shape was placed a little below this disk. This electrode was the electrical equivalent of a vacuum tube grid. The bottom electrode was a small spherical shaped electrode. All electrodes were supported by a central insulating column. The D.C. voltage was applied between the bottom negative electrode and the top positive electrode. The A.C. signal was applied between the "Grid' and the bottom electrode.

The A.C. high voltage signal was in the megahertz range to establish standing wave patterns between the electrodes. Although the lift force was only 100 grams at 150,000 volts, test showed that the lift was exponentially increasing as the 2.5 power of the voltage. Bahnson filed a U.S. Patent in1964 using these principles and it was granted as U.S. Patent number 3,263,102 in 1966.

One point to be made here is the thrust to power ratio of electro propulsion. Some of these devices use very high voltage but very low current so the net power consumed is fairly small. With proper engineering one could build megavolt generators with microamperes leakage currents involved, yielding devices capable of generating hundreds of pounds of force per watt of power consumed. DC 9 jet engines produce about 3 pounds thrust per horsepower or about 0.004 pounds per watt. So, 100/0.004 = 25,000. In other words, electro propulsion could be theoretically 25,000 times more efficient that jet or rocket engines.

Brown also discovered that unsymmetrical electric fields could considerably improve the electrogravitic force and created a 15 inch diameter dome shaped saucer that could lift itself off the ground when 50,000 volts was applied. This was using different geometry and principles than his previous patent. So, he applied for another patent which was awarded as U.S. patent number 3,187,206 on June 1969.

As a matter of fact, unsymmetrical electric fields are very useful at generating unbalanced non-reactive forces. The smaller electrode will have a higher charge density and electric field strength than the larger one. Brown discovered that this configuration would generate a force toward the larger electrode even with the polarity reversed.

Considering that lines of the electric field are always normal to the charged conductor surface and that the attractive force between oppositely charged surfaces is transmitted along these electric field lines of force, another invention comes to mind.

Imagine a horizontal conducting plate with a vertical conducting plate attached in the middle of the horizontal plate, but electrically separated by a dielectric. When the these plates are oppositely charged, the electric lines of force from the vertical plate will originally be going horizontally and bend downward towards the horizontal plate. These lines of force become normal to the horizontal plate in a vertical direction, pulling it upwards. The lines of force on the vertical plate are coming off both sided of it in a horizontal direction, canceling the forces on this plate, resulting in a net upward force on the whole device.

This concept was proven to work by Alexander Frolov, in St. Petersburg, Russia in 2001 at Faraday Labs (now called Faraday Company Ltd.). In fact, a number of interesting inventions and concepts were discussed in their Publication, *New Energy Technologies* Available at: http://www.faraday.ru/

Frolov's electric propulsion experiment was replicated by Jean-Louis Naudin at JLN labs which has another very informative website: http://jnaudin.free.fr/

So, it turns out there are a number of ways to create non-reactive forces (or they are acting on all of space, as some have theorized) on objects using electricity. Other inventors who have discovered some of these ways; J.F. King, U.S. Patent 3,322,374 May 30, 1967, James Woodward, U.S. Patent number 6,098,924, August 8, 2000, Hector Serrano, PCT patent number WO 00/58623, October 5, 2000; Jonathan W. Campbell U.S. patent number 6,317,310, November 13, 2001, U.S. Patent number 6,411,493, January 31,2002, U.S. patent number 6,775,123,

August 10, 2004.

It is interesting that Jonathan W. Campbell works at NASA but NASA doesn't officially seem to be using his ideas in their space exploration. The thrust to horsepower ratio of electric propulsion is orders of magnitude greater than rocket propulsion. Are they just stupid or are they not being totally honest with the public?

As previously stated, action at a distance requires an active medium to transfer forces. One researcher, Paul A. LaViolette, Ph.D., has developed a theory called Sub-quantum Kinetics that yields a good explanation of many of the unexplainable mysteries of current physics, including the actual nature of electric charge and the fields that mysteriously cause action at a distance. His theory has made predictions in astrophysics that have proven to be correct. He also has written *Secrets of Antigravity Propulsion*, a book I highly recommend reading for the more serious person studying antigravity. In this book he thoroughly covers the field of electrogravitics and other topics.

LaViolette also was in touch with Dr. Podkletnov who was investigating superconducting effects on gravity. Podkletnov had developed a gravity impulse beam generator which was described in Jane's Defense Weekly, July 29, 2002 under the title *Antigravity Propulsion Comes out of the Closet* by N. Cook. In 2003, Violette explained his Sub-quantum Kinetics theory to Podkletnov and made several predictions based on his theory. One, there would be zero reactive force on the beam producing equipment; two, the gravitational beam impulse will travel faster than the speed of light.

Dr. Podkletnov set up experiments to test these two predictions. He used a discharge of 2 mega volts from a Marx generator and his gravity impulse beam passed through a Faraday Cage, 30 centimeters of brick wall and 2.5 centimeters of steel and displaced an 18.5 gram weight hanging from an 80 centimeter thread by 14 centimeters. The gravitational impulse beam traveled 150 meters from beam emitter to the pendulum and retained its 10 centimeter diameter for the entire path. He discovered that, indeed there was no reactive force on the beam generator. His measurements of the impulse beam velocity determined that it had a velocity of between 63 and 64 times the speed of light! This pretty much disproves Einstein's Special Theory of Relativity.

Later, in July 2003, Dr. Podkletnov revealed to Paul LaViolette that he had created a more powerful beam using an improved Marx generator that could provide 10 Megavolt pulses with much quicker rise times. These gravitational impulse beams could punch 4 inch diameter holes in concrete blocks and substantially dent half inch steel plate! The measured velocity of these more powerful gravitational pulses was over a thousand times the speed of light!

LaViolette then went on to propose how these gravitational impulse beams could be used for space propulsion in his book. In Chapter 7 & 8, LaViolette discussed a classified project called Project Skyvault and phase conjugate microwaves, used to propel spacecraft, create Star Wars weapons and provide more secure electronic communications. Also he discussed how phase conjugate microwaves could be used as "tractor beams" that could pull objects towards your space ship.

Paul LaViolette started the Starburst Foundation, which does basic research. http://www.starburstfound.org His theory of Sub-quantum Kinetics can be obtained for free at http://www.starburstfound.org/downloads/physics/cosmic-ether.pdf

In 1956, Michael Gladych wrote an article titled *The G-Engines Are Coming!* In the article, it said that gravity research had been supported by Glenn L. Martin Aircraft Company, Bell Aircraft, Lear and several other American Aircraft Companies. It quoted Lawrence D. Bell, "We are already working on nuclear fuels and equipment to cancel out gravity." George S. Trimble, Vice President in charge of the G- Project at Martin Aircraft was quoted, "The conquest of gravity could be done in the time it took to build the first atomic bomb." William P. Lear, Chairman of Lear Inc. was quoted as saying "All matter within the ship would be influenced by the ship's gravitation only, this way no matter how fast you accelerated or changed course, your body would not feel it any more that it now feels the tremendous speed and acceleration of the earth."

Openness about antigravity research continued until 1958 with a number of publications printing stories about the various companies involved. In January, 1958, *Product Engineering* magazine had the following story:

"Electrogravitics: Science or Daydream?

A few weeks from now, at a special session of the Institute of the Aeronautical Sciences (New York City, Jan. 27-31), a group of dedicated men will discuss what some people label pure science-fiction, but others believe is an attainable goal. The subject: electrogravitics-the science of controlling gravity…"

After the meeting, *Business Week* had this to say:

> "If anyone had predicted 10 years ago that a cross-section of the nation's top physicists, aeronautical engineers and mathematicians would be fighting for standing room to hear the chaste theory of gravity seriously challenged, he

would have been labeled sun-stroked, senile, or worse.

...At an opening day meeting of the Institute of Aeronautical Sciences in New York last week, however, the impossible became possible. In record numbers – in a rush that stacked up scientists 20 deep at every entrance to the Sheraton-Astor's North Ballroom – the elite of research came to hear what it is that has reawakened scientific interest in the possibility about doing something about gravitation.

What has happened, they wanted to know, that has caused major aircraft companies as well as the government and various universities, to start serious inquiries into the possibility of controlling gravity? And even more importantly, how accurate are the reports (circulated by *Tass*) that Russian Scientists hope to turn up some sort of machinery to cancel or modify the force of gravity sometime during 1958?..."

By the end of 1958 however, all reportage on gravity research from the companies involved seemed to stop. In the July, 1959 issue of *Canadian Aviation*, Charles Carew wrote, "The author has not been able to determine whether the Glenn L. Martin Corp. has discontinued its anti-gravity program or made a significant discovery which has elevated it to a super-top-secret category, since no information about the project has recently been available."

The same phenomenon occurred with the other companies involved in gravity research. The public would hear no more about the subject which previously was so openly discussed. MJ-12 had decided to place all antigravity research and development under an extreme National Security classification.

At first, Eisenhower was worried about keeping the lid on MJ-12 secrets. On January 24, 1953, Eisenhower established the Advisory Committee on Government Organizations. He hired Nelson Rockefeller to be chairman of the committee. Rockefeller passed on to Eisenhower his recommendations on reorganizing the National Security Agency to protect the secrecy of MJ-12. MJ-12 should be headed by the Director of Central Intelligence. This would insulate MJ-12 from the vagaries of politics of future presidents. Eisenhower agreed to the recommendation.

Later, Eisenhower would regret this decision, as he realized that the corporations which the CIA cleared to work on secretive back engineered alien technology were becoming more powerful than the U.S. government itself and became beyond government control. This was the reason behind his warning about the Military Industrial Complex in his farewell address.

In any case, by 1958, MJ-12 had decided that antigravity research was to be highly classified under National Security and any corporation with military contracts would keep the research and development "black" if they wanted to do further business with the military. Employees working on these black research and development projects were required to sign secrecy contracts which had severe penalties beyond job loss for disclosure of these secrets. Punishments could include up to $10,000 fines and up to ten years in prison.

Officially, these "black projects" are called Special Access Projects (SAP). In the DOD Manual titled: National Industrial Security Program Operating Manual:

"There are two types of SAPs, acknowledged and unacknowledged. An acknowledged SAP is a program which may be openly recognized or known; however, specifics are classified within that SAP. The existence of an unacknowledged SAP, or an unacknowledged portion of an acknowledged program, will not be made known to any person not authorized for this information."

The DOD Manual goes on to clarify the measures taken to keep secret the existence of unacknowledged programs:

> "Unacknowledged SAPs require a significantly greater degree of protection than acknowledged SAPs... A SAP with protective controls that ensures the existence of the Program is not acknowledged, affirmed, or made known to any person not authorized for such information. All aspects (e.g., technical, operational, logistical, etc.) are handled in an unacknowledged manner."

In addition to the stringent security requirements pertaining to an Unacknowledged SAP (USAP), these may be further classified by making them Waived USAPs. According to a 1997 Senate investigation:

> "Among black programs, further distinction is made for "waived" programs, considered to be so sensitive that they are exempt from standard reporting requirements to the Congress. The chairperson, ranking member, and, on occasion, other members and staff of relevant Congressional committees are notified only orally of the existence of these programs."

A 1992 supplement to an earlier version of the DOD Manuel states:

> "Program cover stories. (UNACKNOWLEDGED Program.) Cover stories may be established for unacknowledged programs in order to protect the integrity of

the program from individuals who do not have a need to know. Cover stories must be believable and cannot reveal any information regarding the true nature of the contract. Cover stories for Special Access Programs must have the approval of PSO [Program Security Officer] prior to dissemination."

So, Congress oversight officials have to take the word of the sponsoring military service or intelligence agency that the program is being run responsibly and in accord with U.S. and international laws, but actually have no realistic way of discerning whether they are getting a straight story or a cover story. Since they are informed orally, there is no documentation of who said what to whom and when.

So, realistically speaking, there is no actual Congressional oversight of how taxpayer money is being spent on waived USAPs. Also, CIA "black projects" are financed with CIA "black money" not appropriated through Congress, like money obtained via `transfer from other DOD projects by creative bookkeeping and money laundered from illegal narcotics trafficking and financial fraud. And, as we have seen, the CIA controls MJ-12 and therefore back engineered ET technology and which companies get access to these programs and contracts.

The second category of unacknowledged space activities pertains to private corporations that invoke similar security procedures as the U.S. military or intelligence agencies as a condition for working on highly classified contracts. These industry standard security procedures by the DOD are outlined in the "National Industrial Security Program Operating Manual." Here is how the 1997 U.S. Senate Report summarizes the current situation:

"Industrial contractors performing classified contracts are governed by the National Industrial Security Program (NISP), created in 1993 by Executive Order 12829 to "serve as a single, integrated, cohesive industrial security program to protect classified information." A Supplement to the NISP operating manual (NISPOM) was issued in February 1995 with a "menu of options" from which government program managers can select when establishing standards for contractors involved with special access programs."

One of these options from the "menu of options" was Army-Navy-Air Force publication 146, issued by the Joint Chiefs of Staff in 1953, that made unauthorized release of information concerning UFOs a crime under the Espionage Act, punishable of up to 10 years in prison and a $10,000 fine. Information concerning UFOs could legally include information on back engineered ET technology.

Many of the Aircraft corporations receiving these SAP, USAPs and waived USAPs contracts, which were located around the populous Los Angeles area, started relocating, at least their more secret operations, to the desert regions around Palmdale, California near Edwards Air Force base. Also, alien technology and research was moved from Wright Patterson at Dayton, Ohio, where it could be accessed by U.S. government officials, to Area 51, north east of Las Vegas, Nevada where the CIA could keep things hidden.

"Skunk Works" is an official alias for Lockheed Martin's Advanced Development Programs (ADP), formerly called Lockheed Advanced Development Projects. Skunk Works was located in Burbank, California on the eastern side of Burbank-Glendale-Pasadena Airport. During World War II, there was an odorous plastic factory next door, hence the name Skunk Works. After 1989, Lockheed reorganized its operations and relocated the Skunk Works to Site 10 at U.S. Air Force Plant 42 in Palmdale, California, where it remains in operation today.

A section of Skunk Works was doing much of the classified antigravity research and development. Some working inside corporations like Lockheed Martin resented the fact that antigravity technology being developed was being kept secret and there were the inevitable, occasional, but cautious leaks.

Another interesting corporation is Northrop Grumman. In 1994, Northrop Aircraft merged with Grumman Aerospace to create the company Northrop Grumman. Both companies were previously established in the airplane manufacturing industry, and Grumman was famous for building the Apollo Lunar Module. The new company acquired Westinghouse Electronic Systems in 1996, a major manufacturer of radar systems. Logicon, a defense computer contractor, was added in 1997. Previously, Logicon had acquired Geodynamics Corporation in March 1996 and Syscon Corporation in February 1995.

In 1999, the company acquired Teledyne Ryan, which developed surveillance systems and unmanned aircraft. It also acquired California Microwave, Inc., and Data Procurement Corporation, in the same year. Other entities acquired included Xetron Corporation (1996), Inter-National Research Institute Inc. (1998), Federal Data Corporation (2000), Navia Aviation As (2000), Comptek Research, Inc. (2000), and Sterling Software, Inc. (2000).

In 2001 the company acquired Litton Industries, a shipbuilder and provider of defense electronics systems to the U.S. Navy. Later that year, Newport News Shipbuilding was added to the company. And in 2002, Northrop Grumman acquired TRW, which became the Space Technology sector based in Redondo Beach, CA, and the Mission Systems sector based in Reston, VA, with sole interest in their space systems and laser systems manufacturing

Northrop Grumman and Boeing have also collaborated on a design concept for NASA's upcoming Orion spacecraft (previously the Crew Exploration Vehicle), but that contract went to rival Lockheed Martin on August 31, 2006.

So, we see that Northrop Grumman in addition to being a huge conglomerate of other companies, has access to many electronic, military and aerospace capabilities. They also had a unit working on antigravity research and development. Northrop Grumman has a secret underground facility known as "The Anthill" located in the Tehachapi Mountains, North West of Lancaster, California, where much of this secret work is done.

They developed the B-2 stealth Bomber. Like the workers at the Skunk Works, working on antigravity projects, some at Northrop Grumman also resented all the secrecy. Some secrets of the B-2 stealth bomber started leaking out. *Aviation and Space Technology* magazine ran an article by its editor, Bill Scott stating that the leading edge of the B-2's wing was positively charged with high voltage electricity. Some engineers at Northrop Grumman revealed that negative ions were exhausted from the four General Electric F-118 jet engines to the rear. So, from what we already know, this would create a "gravity well" that the B-2 would keep falling forward into.

These four jet engines can create a total output horsepower of 140,000 horsepower, the equivalent of 25 megawatts. These engines probably also power super conducting generators to provide for the B-2's electrical needs, since they weigh about a tenth of conventional electrical generators and General Electric mass-produces super conducting generators for the Air Force. These generators would easily provide electricity for the flame jet generators that could provide voltages in the millions of volts range.

The wing flaps directly behind these jet exhausts can deflect the negative ion cloud upwards or downwards behind the B-2. In relation to the positive charged leading edge of the B-2 wings, an upward direction of the negative ion cloud would create a downward gravitational vector and a downward direction for the negative ion cloud would create an upward gravitational vector on the B-2. This would increase the maneuverability of the B-2 over just deflecting an uncharged exhaust.

Over at Lockheed Martin's Skunk Works, The secret X-22A, 2 man flying saucer was developed according to Air Force Lt. Col. (retired) Steve Wilson. Pilots are trained to fly this secret craft at a secret aerospace academy at Colorado Springs, Colorado. The X-22A was first test flown at Area 51, Nevada. After passing the tests and getting all the bugs out, missions were then flown out of Beal and Vandenberg Air Force bases in California. These missions regularly fly out into outer space.

According to ex MJ-12 member Dr. Michael Wolf, The Aurora SR-33A can be powered with conventional fuel and antigravity technology and can travel to the Moon. As a matter of fact as we shall later see, there is a U.S. secret military base on the far side of the Moon.

Another plane developed at the Skunk Works is the X-33A Venture Star National Space Plane which can also land and take off from air fields and go into outer space. According to Air

Force Colonial (Retired) Donald Ware, the X-33A has an antigravity system on board. The "A" after the plane design numbers designates Antigravity capability. Because of the high secrecy involved, and resulting confusion, the Venture Star X-33A and the Aurora SR 33-A could actually be the same plane.

Besides using electrogravity, which is the oldest of the antigravity technologies, dating back to the 1920s, there are also other means of creating antigravity. The next level up is magnetogravitics. This involves high energy toroidal magnetic fields which are rotated at high RPMs and sometimes also pulsed. In the 1950s, antigravity researchers called this method "dynamic counterbary."

The secret spacefaring craft called the Nautilus, operated in secret by the Office of Naval Intelligence, used this magnetic pulsing type of propulsion. The Nautilus was developed jointly by the Boeing's Phantom Works and the EU's Airbus Industries Anglo-French Consortium. Boeing's "Phantom Works" is where their highly classified military projects take place. In the past, the Nautilus was used to build the secret base on Mars, operated in conjunction with the Office of Naval Intelligence.

Presently, the Nautilus operates out of the unacknowledged secret headquarters of the U.S. Space Command, located underground, inside of King's Peak in the Wasatch Mountains, 80 miles east of Salt Lake City, Utah. The Nautilus makes twice a week trips to the secret military intelligence space station which is deep in space and jointly operated by the U.S. and Russia for the last thirty years.

An older antigravity space plane, with a 600 foot wing span, is the TR3-B Astra. The triangular shaped nuclear powered aerospace platform was developed under the Top Secret, Aurora

Program with SDI and black budget monies. The Aurora is the most classified aerospace development program in existence and has developed a number of antigravity space craft.

The TR-3B is Code named Astra. The tactical reconnaissance TR-3B first operational flight was in the early 90s. At least 3 of the billion dollar plus TR-3Bs were flying by 1994. The TR-3B was the most exotic vehicle created by the Aurora Program at the time. It is funded and operationally tasked by the National Reconnaissance Office, the NSA, and the CIA.

The TR-3B vehicles outer coating is reactive to electrical Radar stimulation and can change reflectiveness, radar absorptiveness, and color. This polymer skin, when used in conjunction with the TR-3Bs Electronic Counter Measures and, ECCM, can make the vehicle look like a small aircraft, or a flying cylinder–or even trick radar receivers into falsely detecting a variety of aircraft, no aircraft, or several aircraft at various locations. A circular, plasma filled accelerator ring called the Magnetic Field Disrupter (MFD), surrounds the rotatable crew compartment shielding it from gravity and inertia, and is far ahead of any imaginable technology. Sandia and Livermore laboratories developed the reverse engineered MFD technology.

Defense Industry insider, Edgar Rothschild Fouche wrote about the TR-3B in his fact presented as fiction book *Alien Rapture*. This craft has been seen by many. One particular sighting was over Phoenix Arizona, where hundreds of eyewitnesses watched a huge black triangle slowly cross the sky. Other sightings have been in the Hudson River Valley in New York State and other parts of the United States and Europe.

On March 31, 1990, two F16s were scrambled from Beauvechain airbase in Belgian to intercept the, to them unknown, TR3-B, which was traveling at about 150 Knots. When the fighter jets got within 5 to 8 miles of the mysterious craft, they locked-on with their missiles, the TR3-B suddenly picked up speed to 1010 Knots and broke the missile lock-on and flew away. The craft maneuvered in some very high G turns and then suddenly rose 9,500 feet in 5 seconds, estimated at 24 Gs! Needless to say, after that, the F16s could not catch up with the mysterious craft or even again lock on to it with their missiles.

Northrop Grumman developed the B-2 stealth Bomber, which incorporated many of the features mentioned in Thomas Townsend Brown's Winterhaven proposal, in addition to stealth technology.

Another type of propulsion, revealed by whistle blower, Robert Lazar, who worked at the S-4 division of Area 51, uses the gravitational nuclear strong force. To use this force, it must extend past the atomic nucleus which only occurs in elements of atomic number 115 and higher.

The extraterrestrial flying saucer that Robert Lazar worked with used element 115, which at the time of Robert Lazar's revelations in 1989, didn't even exist on Earth. Since then, element 115 has actually been artificially created and is called Ununpentium. According to Dr. Michael Wolf, this saucer and element 115 were supplied to the people at S-4 by the ETs from Zeta Reticuli as part of their technology exchange program.

By amplifying that exposed gravitational nuclear strong force with a gravity wave amplifier, the force could be directed and used to propel the flying saucer. The energy used also came from element 115, which when bombarded with protons becomes element 116 which is unstable and decays, emitting a strong flux of antimatter. The matter – antimatter reaction was efficiently, directly converted into high powered electricity, in a back-engineered device from an extraterrestrial device which did the same thing.

The advanced TAW-50 is a hypersonic, antigravity space fighter-bomber. Developed in the 1990s, in a joint Northrop-Grumman - Lockheed-Martin venture, lots of back-engineered alien technology has been incorporated into the TAW-50. The craft is powered with SCRAM propulsion in the outer atmosphere and can do over Mach 50 or 38,000 MPH. The electrogravitics for the TAW-50 was produced by GE Radionics and Pratt & Whitney designed the SCRAM atmosphere penetration technology.

Using its back engineered alien antigravity technology it can cancel virtually all inertia inside and around the craft and is capable of hair pin turns at 38,000 MPH. It can go from top speed to a complete stop in under 2 milliseconds without hurting the pilots or crew.

The TAW-50 has a crew of 4, but everything happens so fast that all the guidance is computer controlled by a system provided by American Computer Company derived from its Valkyrie XB/9000 Artificial Intelligence Guidance series. It is designed to stay in space for 2 days but could stay longer with extra liquid oxygen carried onboard. In fact with extra LOX, it could fly to the Moon and back. Also, it can refuel at the secret undeclared Military space station in orbit around the Earth that the Space Shuttle used to take occasional classified Defense Department payloads to.

As of 2002, there were 20 TAW-50's in the U.S. arsenal. The military astronauts for these antigravity space planes are trained at a secret Air Force Academy located in the hills just west of the official Air Force Academy at Colorado Springs, Colorado. They rotate duty by flying to and from Vandenberg Air Force base in other antigravity vehicles.

A very interesting covert agency within another covert agency is an unacknowledged department of NSA called the Advanced Contact Intelligence Organization (ACIO) whose primary mission is to research, assimilate and replicate any technologies of extraterrestrial origin. Their well-paid (around $400,000/ year) employees are of extremely high IQ and their technology is advanced beyond any other organization on the planet. The secrecy is such that only a few even at NSA even know of ACIO's existence.

One technology developed by ACIO is called Blank State Technology (BST). BST is a form of time travel technology that allows history to be rewritten.

One extraterrestrial civilization that ACIO became in contact with was the Corteum civilization. They exchanged intelligence acceleration technology which allowed human intelligence to be expanded. The head of ACIO code named "Fifteen" used this technology on himself to boost his IQ even further.

Fifteen was the one who developed the BST technology in conjunction with the Corteum scientists. Fifteen also felt that NSA directors were too immature to handle the BST technology wisely and created another covert group within the ACIO which he called the Labyrinth, originally comprised of himself and members of the Corteum group. Later, other members of ACIO had their IQs boosted by the Corteum technology and were initiated into the Labyrinth. Eventually Fifteen took over total control of the ACIO.

ACIO was also exchanging technology with the Greys but

Fifteen didn't fully trust the Greys who primarily communicated telepathically. Fifteen suspected that the Greys were forcing their agenda too much by the way they dominated the telepathic conversation and possibly they were planting thoughts into human minds that furthered the Grey's agenda. Fifteen kept the BST project secret from the Grey's and NSA. The Grey's considered Fifteen to be the leader of planet Earth because of his high intelligence.

The ACIO had also compiled a vast collection of ancient artifacts and records which were collected for several centuries by a secret organization with present links to ACIO. These records are not known about by mainstream academic society. Many of these records had prophecies of the future history of our planet. These prophecies are what inspired Fifteen to create the BST technologies.

Changing history is a complex operation that requires the ability to correctly forecast the desirable outcome from tampering with the time stream. Different scenarios are run on super computers to determine desirable outcomes, correct insertion points in the space-time continuum, and time required to make the proper changes. The BST is not a toy for immature people with immature self-serving agendas. Also, the persons sent in to make any changes need to have the proper mental requirements to carry out the mission.

In 1972, about 80 miles northeast of Chaco Canyon, some hikers discovered an unusual cave hidden in a remote canyon. Very unusual artifacts and cave pictures were discovered in this cave. Before long, word of the discovery got back to NSA and the ACIO. ACIO sent their own people out to discover what was there and soon the area was placed off limits to the public.

After a while, the ACIO team discovered a 23 interconnected cave complex, all with unusual cave pictures and artifacts. These findings did not fit in with any known native culture and some of the artifacts seemed to be advanced extraterrestrial technology. The ACIO developed a project called Ancient Arrow to learn more about these findings.

In cave 23, an unusual CD like optical disc was discovered among the artifacts. At first, all attempts to decode the information on this disk failed. Finally, clues from the cave paintings were combined with the ancient Sumerian language and a partial translation of the disc was made.

Apparently, the caves were made using sonic technology (perhaps similar to what Keely had developed) by a group that called themselves Wingmakers. The Wingmakers possessed advanced time travel technology and had traveled from 750 years in the future to place a time capsule in place at the Ancient Arrow cave complex in 350 B.C.

Six other time capsules were placed around the planet to be discovered at the appropriate time. The technology discovered there was so advanced that most of it was not even understood by the extraterrestrial races working together with the ACIO to discover how it functioned.

ACIO could only let their scientists work on the Wingmaker, Ancient Arrow material for a number of months before replacing them with new scientists entirely. Apparently, the Wingmakers meant for their subject matter to be known about by all the people of the Earth and by working with their material, one became highly motivated to share the knowledge and would become a security risk. Apparently one of those scientists dropped out of the ACIO and became a whistleblower and that is how the knowledge became known to the public.

This is only a short summary of this interesting subject. More interesting information about this story can be found by placing the word combination "wingmakers ancient arrow project" into an internet search engine. Also Chapter 7 of *The Time Travel Handbook* edited by David hatcher Childress, has part of the story.

Getting back to the subject of advanced black projects by the military industrial complex, Ben Rich was a secret Israeli Mossad agent and was also the former director for Lockheed-Martin's Skunk Works. So, you can bet that the Israeli Mossad had access to most of the secrets of the Skunk Works.

In 1986, Ben Rich wrote, "Many of our manmade UFOs are Un Funded Opportunities." This indicated that some UFOs were produced here on Earth, probably by the Skunk Works, from unfunded black budgets.

At an alumni engineering meeting at UCLA in March 1993, Ben Rich said, while showing a slide with a black disk headed for space," We now have the technology to take ET home. No, it won't take someone's lifetime to do it. There is an error in the equations. We know what it is. We now have the capability to travel to the stars. First, you have to understand that we will not get to the stars using chemical propulsion. Second, we have to find out where Einstein went wrong."

When Ben Rich was asked how the UFO propulsion worked, he said "Let me ask you. How does ESP work?" The questioner responded with, "All points in time and space are connected?" Rich then said, "That's how it works!"

And another time, Ben Rich said, "We already have the means

to travel among the stars, but these technologies are locked up in black projects, and it will take an act of God to ever get them out to benefit humanity. Anything you can imagine, we already know how to do."

On his death bed, Ben Rich confessed that the U.S. military can now travel to the stars and that extraterrestrial visitors to the Earth are real.

8

The *Solar Warden* Fleet

According to their web site, the United States Space Command (USSPACECOM) is a Unified Combatant Command of the United States Department of Defense, created in 1985 to help institutionalize the use of outer space by the United States Armed Forces.

The U.S. Space Command is headquartered at Peterson Air Force base at Colorado Springs, Colorado at the following address and Phone number:

Air Force Space Command

Public Affairs Office

150 Vandenberg St., Suite 1105

Peterson AFB, CO 80914-4500

(719) 554-3731 or DSN 692-3731

The U.S. Space Command had its computer database hacked in 2002, by British Citizen, Gary McKinnon, who discovered on the database a list of "Non-Terrestrial Officers" and " fleet to fleet transfers." The names of the ships listed on the "fleet to fleet transfers" did not correspond to any U.S. Navy ships.

McKinnon also stated that the records revealed that the off-planet shuttles could hold 300 people. That statement is reinforced by a statement from President Ronald Regan's diary. This diary entry for Tuesday June 11, 1985 reads:

"I lunch with 5 top space scientists. It was fascinating. Space is truly the last frontier and some of the developments there in astronomy etc. are like science fiction, except they are real. I learned that our shuttle capacity is such that we could orbit 300 people."

The space shuttle that the public knew about could only hold 8 people. So obviously there was one that was secret that Reagan knew about and McKinnon discovered that could hold 300 people. And you can bet that it used antigravity technology.

Reagan also made an interesting speech at the 1985 U.N. General Assembly:

> "In our obsession with antagonisms of the moment, we often forget how much unites all members of humanity. Perhaps we need some outside, universal threat to make us recognize this common bond. I occasionally think how quickly our differences worldwide would vanish if we were facing an alien threat from outside this world. And yet I ask- is not an alien force already among us?"

This reference to aliens was not on the speech his staffers had approved. So, Reagan ad libbed this speech. And, what did he mean about an alien force already among us? Interestingly

Gorbachev responded to this speech at the 1985 Geneva Summit:

> "At our meeting in Geneva, the U.S. president said that if the earth faced an invasion by extraterrestrials, the United States and the Soviet Union would join forces to repel such an invasion. I shall not dispute the hypothesis, although I think it's early yet to worry about such an intrusion."

But, from many other sources it appears that the U.S. and Soviets were already cooperating in Space. Perhaps the Cold War, like NASA was also a smoke screen.

Does McKinnon's discovery mean the Space Command has fleets of space ships in space that Non-Terrestrial Officers get transferred between in fleet to fleet transfers? And, more importantly, when do these Non-Terrestrial Officers get " shore leave" back on Earth?

McKinnon also hacked into NASA and Pentagon computers where he found film, photographs and other evidence of alien spacecraft secretly held by these agencies. One picture in space showed a high definition photograph of a large cigar shaped object over the northern hemisphere of planet Earth.

The Bush Administration accused McKinnon with a serious breach of U.S. national Security. McKinnon would have faced a possible 70 year prison sentence. The U.S. tried to extradite him to the U.S. for trial but failed. Some think that the U.S. government really didn't try too hard to extradite McKinnon and bring him to trail. A trial would have created much more publicity to the U.S. secret space program and could possible create an airing of more secrets.

Another clue that the Space Command has secret space craft, not known about by the public, was the August 6, 2007 award that NASA gave "The Human Spaceflight Support Team" which was part of the U.S. Space Command, for assisting NASA.

Part of the awarding citation says,"... highlights the team for its exceptional support in keeping the Space Shuttle, International Space Station and its crews safe from the dangers of orbital debris, spacecraft collisions and other inherent hazards of orbit operations."

So, how is the U.S. Space Command able to clear this orbital debris, unless they have their own space craft with weapons or other devices (tractor beams?) capable of destroying or removing the debris?

Another Space Command (possible in conjunction with the Air Force Space Command) code named "Solar Warden" is operated by the US Naval Network and Space Operations Command (NNSOC) [formerly Naval Space Command], headquartered in Dahlgren, Virginia.

The evolution began in September 2005 with the integration of the Naval Security Group (NSG) and its subordinate commands into the Naval Network Warfare Command (NETWARCOM), Norfolk, Virginia. It continued unfolding in September 2006 when the Naval Network and Space Operations Command (NNSOC), Dahlgren, Virginia, was disestablished, and NETWARCOM subsumed its responsibilities as well. Many of the NNSOC's 250 billets are being used to staff NETWARCOM's Network Information Operations and Space Center.

Since the Naval Network Warfare Command took over NNSOC they are now in charge. But , MJ9 Commander John Chandeler is still in charge of the Solar Warden space fleet. At the time, these "Ships" consisted of 8 cigars shaped "Mother ships" longer than 600 feet acting as space born aircraft carriers for the 43 smaller scout ships, according to Dr. Richard Boylan.

Dr. Boylan started out as a behavioral scientist who worked with persons having ET encounters. After hearing many cases, he started giving talks at UFO conferences. This attracted the attention of certain people working on black projects who wished to make some information public without risk to themselves. These people started feeding information to Dr. Boylan. This is the source of his information. Much of this information correlates with information from other sources.

One Mother ship was lost recently to an accident in Mars' orbit while it was attempting to re-supply the multinational colony on Mars. This base was established in 1964 by American and Soviet teamwork.

The first joint U.S. Soviet mission to Mars was in May, 1962 according to information obtained by researcher Vladimir Terzerski. So, the "cold war" was a smokescreen for public consumption, while secretly, the U.S. and Soviets were working jointly on a secret space program. This illusion also gave the excuse to the public for a "space race" to spend billions on NASAs space program. Some of that money was funneled off into the secret space program through the Unacknowledged Special Access Project programs gambit.

The public space program via NASA is merely a "smoke screens" to hide the real but secret space program. The Nazi paperclip scientists brought to the U.S. after World War II, included Dr. W.O. Schumann, the father of the German flying saucer projects. Why waste your time and money on rockets to get into outer space when you already have antigravity technology?

Civilian researchers who successfully created antigravity technologies were effectively suppressed, as occurred to Otis Carr in 1961. This suppression was revealed for the first time in 2007 through the testimony of Carr's former protege, Ralph Ring. Another suppression of John Searle's levity disk occurred in England. And Howard Menger's Electrocraft was forbidden by the FBI on National Security grounds.

As Area 51 was becoming well known about, another secret hardened underground facility was built inside of King's Peak in the Wasatch Mountains, 80 miles east of Salt Lake City, Utah. This is the new U.S. Space Warfare Headquarters. The X-22A antigravity saucers, equipped with neutral particle beam weapons, among other craft, are deployed from this underground base.

Also, a huge triangular craft, estimated to be 600 feet long by 100 feet across, was seen by a hiker in a remote region of the high Utah dessert. The huge craft was silently hovering and gradually losing altitude. Then, the dessert floor opened up. Or rather, some huge doors camouflaged to look like dessert opened up. The craft lowered itself into the opening and the camouflaged doors swung close and the area looked just like dessert again.

So apparently, there was an underground base right there in the middle of the desert. And this base could launch and land these huge air/space craft. It is estimated that a craft that size could carry up to 2,000 passengers.

The history of the military space program started right after World War II with the help of the Paperclip Nazis. First, were the intercontinental ballistic missiles developed from the Nazi V-2 rockets at White Sands, New Mexico. Then, came the multiple stage rockets that could place military spy satellites into orbit.

Parallel with these known projects were the more secret programs. There was the Dyna-Soar winged space plane to take military astronauts into outer space that could take off and land at special air fields. This space plane, originally designed by Nazi Scientist, Eugen Sanger, was the precursor to the space shuttle. The important difference was that the Dyna-Soar was able to take off without a booster rocket by using electrogravitics. The military trained a military astronaut corps to pilot the Dyna-Soars. The official first man to walk on the Moon, Neil Armstrong, was an ex Dyna-Soar pilot.

Also, there was the secret development of the German flying saucers into improved U.S. flying saucers.

The Military was also involved with the secret Manned Orbiting Laboratory (MOL) project and from 1965 to 1967 trained a cadre of astronauts for this project.

Also, there was the military Project Horizon which planned to build a base on the moon.

Officially, Dyna-Soar, MOL and Project Horizon were all shut down. Secretly, they are all operational with even better equipment and more ambitious plans, as we shall see.

The history of Solar Warden started in 1962, with secret meetings between the Star Nations and 4 major Earth Governments, The U.S., Russia, China, and the European Council (for runner to the European Union). At that time MJ-12 was internationalized under the United Nations Security Council and the original MJ-12 became the U.S. Special Studies Group. That way, the U.S. wouldn't have exclusive control over ET related matters.

The current members of this international MJ-12 are:

MJ1 - Simonetta Di Pippo, the Italian director of UNOOSA.

MJ2 – Brigadier General Gartzene Killennarry, senior manager at British MI-6.

MJ3 - Anke Schaferkort, German director of BASF.

MJ4 – Hon. Paul Hellyer, former Canadian Minister of Defense.

MJ5 – Samantha Power, US Ambassador to the UN.

MJ6 – Thai Male, Professor of Software Engineering at Bangkok University's School of Science and Technology.

MJ7 – Nelson Violante Carvalho, Ph.D. tenured lecturer at the Federal University of Rio de Janerio.

MJ8 – Olga Golodets, Deputy Chairman for Social Affairs , Federal Cabinet of the Russian Federation.

MJ9 – John Chandler, Commander of the Space and Naval Warfare Systems Command.

MJ10 – Joyce Victoria Bigio, Italian Chief Liaison to the Vatican.

MJ11 – Carmen Omonte Durand, Peruvian Congresswoman

MJ12 – Dr. Ross McKenzie, professor of Quantum Physics at University of Queensland.

MJ-12 has been directed in recent times by an Executive Committee composed of CEO Simonetta Di Pippo(MJ1), Samantha Power (MJ5), Joyce Victoria Bigio (MJ10), and Captain John Chandler, USN (MJ9).

Named Director of UN Office of Outer Space Affairs (UNOOSA) in March, 2014, Simonetta Di Pippo is overseeing the gradual integration of MJ-12's Executive Committee into UNOOSA as an executive advisory committee. This is being done because soon there will be no need for a UFO secrecy-management organization.

After integration, members of MJ-12 will be converted into advisors for UNOOSA as needed. Ms. Di Pippo is working with Dr. John Ashe, UN President, to increase the democracy of UN government by strengthening the role of THE GENERAL Assembly and the UN President, and revising the obsolete, oligarchical arrangement of the SECURITY COUNCIL's making all the major decisions. (11)

In 1964, a secret meeting took place in an underground base under El Capitan Peak in northwest Texas. The meeting was the cumulation of a series of meetings first initiated by President Eisenhower in 1960. The U.S. was represented by David Rockefeller, the USSR by Nikita Khrushev, China by a trusted associate of Mao Tse-Tung and the European Council by Anthony Bradley. The Star Nations representative was Asheoma Maeta.

The Star Nations concerns were the irresponsible testing of nuclear weapons and the great harm to the Earth these tests were causing. The nuclear explosions also were causing tears in the space time continuum and were causing problems for the beings on other dimensions. Another concern was the shooting down of their craft by the military of different Earth governments using their new high energy pulsed radar systems.

The 4 Earth Nations representatives wanted the Star Nations to give them more ET technology including antigravity and genetic technology in return for a promise of a cease fire. They also wanted the ETs to restrict how close their mother ships could approach the Earth to stay out of view of the Earth people. The scout craft were limited to how close they could fly to the Earth unless on a mission pre-announced to the government. They also wanted the ETs to not interfere in any of the Earth's religions or commercial arraignments. Finally a treaty was agreed to and signed in 1964. After this, private contactee meetings between private people and ETs became considerably reduced.

After decades of violations of the treaty between the U.N. and the Star Nations, the worst being the shooting down of Star Nation space craft, the 1964 treaty was deemed null and void by the Star Nations.

Finally, it was decided that a certain faction was responsible for these treaty violations. This faction was identified as controlled by the Jesuit order of the Vatican, whose secret societies had infiltrated the banking, energy, military and political segments of many societies. They feared the ETs because they could threaten their economic, military and political control of the world's people. This faction became known as "The Cabal". Known leaders of this Cabal include the Rockefellers, the Rothschilds, Henry Kissinger, Prince Bernhard of the Netherlands, George Herbert Walker Bush, Pope Pius XII and many more of similar ilk who have caused so many unnecessary bloody and cruel wars on our planet.

Finally, negotiations were continued with the understanding that the government leaders truly did not truly represent the will of the people of Earth. The UN Committee on the Peaceful Uses of outer Space was transformed into the Outer Space Affairs Division in the UN Security Council in 1968. It was transformed again into The United Nations Office for Outer Space Affairs (UNOOSA) in 1993, whose stated purpose is "promoting international cooperation in the peaceful use of outer space." At that time, the office was relocated to the UN office in Vienna, Austria.

Solar Warden is part of a secret extraterrestrial treaty agreement with the Star Nations – the organization of advanced intelligent civilizations in space – and the United nations. Because of its advanced technological position, the U.S. has been designated by Star Nations to a lead position in providing space security for Earth. The U.S has been adding to the Solar Warden fleet since the late 1980s using black projects within the aerospace corporations.

But, the U.S. is not the only nation involved. Other nations with advanced technologies are also involved including Canada, United Kingdom, Italy, Austria, Russia, and Australia. Also, there is a U.N. authority involved, the United Nations Office for Outer Space Affairs (UNOOSA).

While the majority of the people staffing the mother ships and scout ships of the Solar Warden Space Fleet are Americans (U.S. Naval Space Cadre), there are also some crew members from UK, Italy, Canada, Russia, Austria, and Australia.

The mission of Solar Warden is twofold:

One part of the Space Fleet's mission is to prevent rogue countries or terrorist groups from using space from which to conduct warfare against other countries or within-country targets. Star

Nations has made it quite clear that space is to be used for peaceful purposes only.

The second part of the Space Fleet's mission is to prevent the global-elite control group, the Cabal, from using its orbital weapons systems, including directed-energy beam weapons, to intimidate or attack anyone or any group it wished to bend to its will.

Because the Space Fleet has the job of being Space Policeman within our solar system, its program has been named Solar Warden. Star Nations has not given the U.S. Government exclusive authority to police the Earth. The U.S. has no authority from Star Nations to engage in any international policing activities. Star Nations has the policy position that the citizens of Earth have the responsibility to work out the operation and regulation of their societies as best they can. The mandate and jurisdiction of the Solar Warden Space Fleet is space. It does not have jurisdiction and does not meddle in Human affairs on the ground, nor Human activity occurring within Earth's atmosphere. Those are the jurisdictions of the respective governments in each country and the air space above their territories.

The Solar Warden Space Fleet's vessels are staffed by Naval Space Cadre officers, whose training has earned them the prestigious 6206-P Space Operations specialty designation, after they have graduated from advanced education at the Naval Postgraduate School in Monterey, California and earned a Master of Science degree in Space Systems Operations.

Secretly, UNOOSA can call on the UN's Central Security Service (UN-CSS) to coordinate with the Solar Warden space fleet.

There are 2 major secret commands under UN-CSS:

1. Space and Naval Warfare Division under MJ9 Commander John Chandeler

2. Special Operations Command

Neither will appear on the UNOOSA's website because they are classified: http://www.oosa.unvienna.org/

From their thirty-fourth session (May12, 2014), UNOOSA's provisional agenda includes:

"Coordination of future plans and programs of common interest for cooperation and exchange of views on current activities in the practical applications of space technology and related areas."

"Contributions of space based technology for climate change adaptation and mitigation."

"Use of space-based technology for disaster risk reduction and emergency response."

"Use of spatial data and activities related to the United Nations Geographic Information Working group and the United Nations system: directions and anticipated results for the period 2014 -2015."

"Preparation of a special report on initiatives and applications for space related inter-agency cooperation."

"Means of strengthening the role of the Inter-Agency Meeting on Outer Space Activities."

"Other matters."

Another source, describing Solar Warden is an anonymous whistleblower known as 'Henry Deacon' who works at Laurence Livermore laboratories as a physicist. Deacon's true identity is known to the creators of the Project Camelot website who have witnessed his credentials and believe him to be credible.

Project Camelot is a whistle blower interviewing organization whose mission is to "Uncover the truth, one whistle blower at a time." Their website is: http://projectcamelotportal.com

According to Project Camelot, "Henry confirmed the existence of a large manned base on Mars, supplied through an alternative space fleet codename Solar Warden." Later on, Henry Deacon revealed that his true name was Arthur Neumann.

Another Project Camelot whistle blower was former Air Force Captain Mark Richards who was also involved with the Dulce Wars and the Secret Space Program. Mark Richards father, Ellis Loyd Richards, Jr. had met the ETs at Holloman Air Force Base in 1964 and was later to order the attack on the Dulce base in 1979.

Mark Richards stated that the U.S. military works with the Reptiod race of the Reptilians on joint projects. There are a number of different reptilian races some allied with and some enemies of the Earth. He also confirmed the Mars Bases (there is more than one) and Solar Warden which he clarified was to defend the inner part of the solar system from unwanted invasion.

He also stated that there is an underground Reptilian base under the Vatican and that they control the Vatican via telepathic hypnosis. These Reptilians have been here for thousands of years, hiding in their underground bases, and consider the earth theirs.

So, even though Solar Warden and the Earth Defense Force is highly classified, information about it is leaking out from a number of different sources that makes it possible to form a picture of what it is all about.

9

Beyond Anti-Gravity

Well, all this technology seems pretty fantastic and seems like science fiction. But, hang on - there is more.

What about Billy Meir's claim to have traveled in time and another universe on the Plejaren space craft? Is it possible to actually travel in time? Is there another Universe besides this one? Let's see what some have to say about these issues.

But first, I have to state a disclaimer. My background is in Electrical Engineering, Physics and free energy systems. Much of the material I have presented to this point, I did so with a high certainty in the working principles involved. As far as the true nature of time or multidimensional space goes, I do not claim to have expertise. So, I can only relay what others have either told me in person or I have read about from others.

In the mid-1980s to the late 1990s, I attended numerous UFO and "New Age" workshops in Los Angeles, California, Sedona, Arizona and Colorado Springs, Colorado. One quite interesting person that I met in all three areas was Alfred Bielek, who I have seen in five different Workshops and have talked with him several times.

Alfred Bielek claims to have been involved in both the Philadelphia Experiment and the Montauk project. The Montauk Project, headed by John Von Newman, essentially got the bugs worked out of the Philadelphia Experiment and perfected both teleportation and time travel by 1983, according to Alfred Bielek. Bielek also confirmed that Thomas Townsend Brown, Tesla and Einstein, among others, worked on the Philadelphia Experiment. Einstein had actually completed his unified field theory at this time but since it was being used in Project Rainbow it was classified as a military secret.

Even stranger, Al Bielek claims that he started out as Edward Cameron but on August 12, 1953 he was sent back to 1927 and started life again as Alfred Bielek. So apparently time travel was operational in 1953. The August 12 date was important according to Al Bielek because an Earth biorithym occurs every 20 years peaking on August 12 with minor peaks every 10 years. 1943, 1963, 1983, 2003 and 2023 would be peak years and 1953 would be a secondary (lesser) peak year. The time portals seem to work better on these years.

Ed Cameron went to Princeton, where he met Dr. John Von Neumann. He later received a PhD at Harvard. Then in 1939, Ed and his brother Duncan Cameron were recruited to work for the Navy. Von Neumann recruited both brothers to work on Project Rainbow. But first they had to relearn physics as Von Neumann taught them how gravity, quantum physics and time really work. Ed Cameron had to learn the physics behind invisibility to give progress reports to the Navy.

The brothers were assigned to the USS Pennsylvania and were scheduled to leave for Pearl Harbor on December 5, 1941. But their orders were cancelled because it was known that Japan would attack Pearl Harbor and both men were too valuable to sacrifice. Later, they were assigned to the Destroyer Escort Eldridge.

Both brothers were on the Eldridge while the ship disappeared and became frightened and jumped overboard. Apparently the ship was in hyper space and both brothers landed in the future in 1983 at Ft. Hero, Long Island. There, they met a 40 year older Von Neumann who explained to them that their ship was lost in hyperspace and that they had to be sent back to the Eldridge to shut down the power to the invisibility electronics for the ship to reappear in normal space.

They were sent 40 years back in time back to the Eldridge and after much effort, which included cutting power cables with fire axes, were able to turn off the ships invisibility electronics and the Eldrige returned to the normal world. However, this experiment had opened a rift in space time and a major time loop between 1943 and 1983 which created other problems.

Ed Cameron was later sent to Los Alamos national laboratory to work with Edward Teller on the Hydrogen bomb project. Ed Cameron left this project after some disagreements with Edward Teller in 1947.

Later Ed Cameron worked with Jack Ridley on an ion rocket drive system for use in space which developed a successful prototype in 1953 which could develop 1,200 Lbs. of thrust for 20 minutes. His father told them that they should start a company because this would be the technology of the future. He also offered to finance the company.

So they formed the California Company, JRC Enterprises. But apparently, someone powerful didn't want the ion rocket engine developed. Edward Cameron suspected it was the Cristaldi

Research Group.

Before long, a group of black operations soldiers forcibly removed Edward Cameron from his office and placed him on a train for the Pentagon. From there Ed was taken to Ft. MacLean, VA. There was a teleportation portal there, which was used to send Ed Cameron to a planet orbiting Alpha Centauri One.

On the planet of Alpha Centauri One, Edward was thoroughly interrogated about many of the unusual parts of his life by the semi human aliens. Edward figured that he better tell the truth or he may never get back to Earth. At one point Edward Cameron asked them if they knew about the Cristaldi Research Group. The aliens said "Oh yes, we operate it." After the several day interrogation was completed, Edward was sent back to Ft. MacLean.

After that, his life was in limbo. He was at the Pentagon and asked the Joint Chiefs what his assignment would be. They said that they did not know. Finally, he got to speak with the Commander of the Joint Chiefs. Edward explained that he needed a project to work on and without one his life would be a waste. The Commander's eyes teared up and replied that the matter was beyond his control.

Finally on August 12, 1953, Edward was returned to Montauk, age regressed and time traveled back to 1927 to start his new life as Alfred Bielek.

When I first heard Alfred Bielek's story, I thought "His story is so incredible that it has to be true. Nobody would make up a story like that and expect it to be believed." But, one has to wonder just when time travel actually was perfected. Evidence that I have seen causes me to believe that Tesla was the first to experiment with it and it was perfected in Project Phoenix, which was started after Project Rainbow and the "Philadelphia Experiment."

Other persons involved in the Montauk project, at Fort Hero, Long Island, were Duncan Cameron and Preston Nichols. Peter Moon and Preston Nichols co-authored the book *Montauk Project* which explains some of what happened there. I also got to see Preston Nichols, who with Alfred Bielek, conducted a Delta T (change in time) workshop at the International Tesla Society in Colorado Springs in 1993.

At this conference, I also saw, for the first time, a working free energy machine presented by Joseph Newman. Before that I was open minded but skeptical (because of my formal Electrical Engineering schooling which taught that it was impossible). After seeing Newman's free energy machine, I became a believer.

So, a number of people who worked on the Montauk project claim that teleportation and time travel technology has been manufactured and perfected. Alfred Bielek claims that the Mars colony was having problems entering an ancient pyramid on Mars. So, a teleportation beam was dialed in from Earth to an inner chamber in the Mars pyramid and Bielek himself was teleported inside this pyramid on another planet!

Another person that I exchanged information with was UFO researcher Vladimir Terziski, a scientist from the Russian Academy of Science, whom I have met with several times in Los Angeles and Sedona. He had done a lot of research on Nazi flying saucers in Eastern Europe and gave me some videos. One showed numerous pictures of Nazi flying saucers. The other one about *Alternative Three*, showed a film clip of the joint U.S. - Russian Mars landing on May 22, 1962 in an antigravity flying saucer.

In the background of the film clip, one could hear excited voices in both Russian and English. The information on the video showed the air temperature and pressure on Mars was comparable to that on Earth on a high mountain.

Alternative Three is the third alternative for surviving a terminal Earth disaster – placing human colonies on other planets. *Alternative Two* was placing humans in underground cities. *Alternative One* was blowing a hole in the ionosphere with a nuclear device to release pollutants. *Alternative One* was a stupid idea. The other two are being actively pursued.

At one conference in Sedona, sponsored by Helga Morrow, the turnout was only two, my lady friend Zhenya Rice and I, because the people at the front desk of the Hotel, where the conference was to be held, didn't know about it (or the conference was purposely sabotaged) and turned people away. Having met Vladimir Terziski earlier that day, I knew all about the conference and therefore was present.

The conference was a flop. But Alfred Bielek, Vladimir Terziski, Helga Morrow, myself and my lady friend; all became well acquainted while waiting for customers that never showed up.

Zhenya Rice was from the Ukraine and spoke fluent Russian with Vladimir Teriski and they became good friends. Vladimir was operating on a shoestring, driving his Volvo station wagon packed full of his publications and VHS videos. Zhenya Rice invited Vladimir to stay at her guest room. Now days, Vladimir Teriski's DVDs are available at Amazon.com.

Helga Morrow agreed to join Al Bielek on an expedition to Montauk with Preston Nichols and Duncan Cameron and later, they all ended up being in the book: *Montauk Revisited:*

Adventures in Synchronicity by Peter Moon and Preston Nichols. On that adventure, Helga learned that her father, a German scientist who was always very secretive with her about his work, also had worked on the Montauk project.

I would invite any sceptics about the Montauk Project to examine all the reference material before closing their minds on this subject. A number of people have come forward stating that they were also involved with the secret Montauk Project at Fort Hero after it was officially decommissioned.

Another whistle blower, Seattle-attorney Andrew Basiago claims as a child, he was a participant in Project Pegasus in the late 1960s and early 1970s which was the US time-space exploration program at the time of the emergence of time travel in the US defense-technical community.

He is a prominent figure in the Truth Movement leading a campaign to lobby the US government to disclose such truths as the fact that the US has achieved "quantum access" to past and future events and has used teleportation to place a secret US presence on Mars.

Andrew Basiago also claims to have traveled in time. His father was involved in Project Pegasus. The chronauts, as they were called, had greater psychological problems with adults whose reality program was already firmly established in their minds, than with children, whose minds were more flexible. So a chronaut training program was established for young people.

As an eleven year old child, Andrew Basiago was sent back in time to President Lincoln's Gettysburg Address. He even has found an old photograph of that time that he is in, as proof.

Andrew Basiago has said the following about his present goals:

> "Imagine a world in which one could jump through Grand Central Teleport in New York City, travel through a tunnel in time-space, and emerge several seconds later at Union Teleport in Los Angeles. Such a world has been possible since 1967-68, when teleportation was first achieved by DARPA's Project Pegasus, only to be suppressed ever since as a secret weapon. When my quest, Project Pegasus, succeeds, such a world will emerge, and human beings linked by teleportation around the globe will proclaim that the Time-Space Age has begun!"

His Truth Campaign about time travel and life on Mars is based on direct, personal experience serving on two US defense projects. He was called back into government service in the early 1980's, when he made numerous visits to Mars after being tapped to join the CIA's Mars "jump room" program.

Andrew Basiago was the editor of Alfred Lambremont Webre's book, *Exopolitics: Politics, Government, and Law in the Universe* (Universe Books, 2005), which uses as a case study human contact with an advanced civilization on Mars. He has a website at: http://www.projectpegasus.net/andrew_d_basiago

Other whistle blowers have come forth about secret Teleportation "Jump Rooms" between El Segundo, California and the Mars colony. William Stillings has come forth and stated that he was involved with the Mars Jump room and stated:

> "I can confirm that Andrew D. Basiago and Barack Obama (then using the name "Barry Soetoro") were in my Mars training course in Summer 1980 and that

during the time period 1981 to 1983, I encountered Andy, Courtney M. Hunt of the CIA, and other Americans on the surface of Mars after reaching Mars via the jump room in El Segundo."

According to Mr. Basiago and Mr. Stillings, in summer 1980 they attended a three-week factual seminar about Mars to prepare them for trips that were then later taken to Mars via teleportation. The course was taught by remote viewing pioneer, Major Ed Dames, who was then serving as a scientific and technical intelligence officer for the U.S. Army. It was held at The College of the Siskiyous, a small college near Mt. Shasta in California.

They state that ten teenagers were enrolled in the Mars training program. In addition to Basiago and Stillings, two of the eight other teenagers in Major Dames' class that they can identify today were Barack Obama, who was then using the name "Barry Soetoro," and Regina Dugan, who Mr. Obama appointed the 19th director and first female director of the Defense Advanced Research Projects Agency (DARPA) in 2009.

Mr. Basiago was the son of an engineer for The Ralph M. Parsons Company, which is headquartered in Pasadena. Mr. Stillings was residing in La Canada, California, which is a suburb of Pasadena. Mr. Obama had just completed a year of undergraduate studies at Occidental College in Eagle Rock, California, near Pasadena. Ms. Dugan was attending the California Institute of Technology, which is located in Pasadena.

The Ralph M. Parsons Company is one of the three dominant, world-wide, aerospace engineering and construction firms in America. According to Robert W. Beckwith, a consulting engineer to DARPAs Project Pegasus, the Ralph M. Parsons Company was involved with Project Pegasus. Beckwith also confirmed that the works of Nicola Tesla allowed for "levitated teleportation" and the Pegasus Project.

As many as seven parents of the ten students, all with ties to the intelligence community, audited the class. They included Raymond F. Basiago, an engineer for The Ralph M. Parsons Company who was the chief technical liaison between Parsons and the CIA on Tesla-based teleportation; Thomas Stillings, an operations analyst for the Lockheed Corporation who had served with the Office of Naval Intelligence; and Mr. Obama's mother, Stanley Ann Dunham, who carried out assignments for the CIA in Indonesia.

From 1981 to 1983, the young attendees then went on to teleport to Mars via a "jump room" located in a building occupied by Hughes Aircraft at 999 N. Sepulveda Boulevard in El Segundo, California, adjacent to the Los Angeles International Airport (LAX).

The actual jump room was like going into an elevator with doors on both sides. Once in the jump room, the space inside would seem to be twisted in a gut wrenching experience. The rectangular room seemed to be twisted into a Cylinder. After a few minutes, the room would untwist back into a rectangular shape. Then, the other door would open and one would walk out on to Mars.

It seemed that Howard Hughes himself was involved with the Mars project. His secretary stated that once Howard had some strange hardware on his desk and instructed her that he was going out but later, some men from Mars were going to come by this office and that she was to give these men the objects on his desk. They later came by and she gave them the hardware.

Basiago said that the time travel and teleportation used in the late 1960s and early 1970s was much gentler than the Mars Jump Room technology. Perhaps that was because the former was based on Tesla's work and the latter on Grey ET technology. Perhaps the Tesla technology wouldn't work as far away as Mars.

Mr. Basiago has also publicly confirmed that in 1970, in the company of his late father, Raymond F. Basiago, he met three Martian astronauts at the Curtiss-Wright Aeronautical Company facility in Wood Ridge, New Jersey while the Martians were there on a liaison mission to Earth and meeting with US defense-technical personnel.

Basiago, Obama, Stillings, and Dugan went to Mars at a time when the U.S. presence on Mars was only just beginning but many had already gone before them.

Mr. Basiago states that in the early 1980's, when they went, the U.S. facilities on Mars were rudimentary and resembled the construction phase of a rural mining project. While there was some infrastructure supporting the jump rooms on Mars, there were no base-like buildings like the U.S. base on Mars first revealed publicly by Command Sgt. Major Robert Dean at the European Exopolitics Summit in Barcelona, Spain in 2009.

The primitive conditions that they encountered on Mars might explain the high level of danger involved. Mr. Basiago and Mr. Stillings agree that Major Dames stated during their training class at The College of the Siskiyous in 1980: "Of the 97,000 individuals that we have thus far sent to Mars, only 7,000 have survived there after five years."

In light of these risks, prior to going to Mars, Mr. Basiago received additional training from Mr. Courtney M. Hunt, a career CIA officer.

"When I asked Courtney M. Hunt why I had to teleport to Mars, he said, 'Because the survival of the human race depends on it.' At this point, we can only guess at what my CIA contact knew. I wish I knew."

"Clearly, I was qualified to teleport to Mars because I had teleported back and forth between New Jersey and New Mexico as one of the children attached to Project Pegasus. But the broader reason why I had to teleport to Mars remains a mystery. "

"Maybe it involved the normalization of relations between Earth and Mars pertaining to an extraterrestrial defense posture that was then being established. Maybe it was something as basic as the fact that they knew that my discovery of life on Mars would inspire the next generation of Americans to become astronauts and space scientists."

"All I know is that it was a journey to Mars that the CIA had determined that I absolutely had to make and that I could not decline. Today, I'm glad that I went to Mars, because it gave me an understanding of what Mars is like, and this understanding informs the works of natural history that I am currently writing about life on Mars."

Hunt showed Mr. Basiago how to operate the respiration device that he would wear only during his first jump to Mars in July 1981, provided him with a weapon to protect himself on Mars, and took him to the Lockheed facility in Burbank, California for training in avoiding predators on the Martian surface.

When they then first teleported to Mars in Summer 1981, the

young Mars visitors confronted the situation that Major Dames had covered at length during the class the previous summer – that one of their principal concerns on Mars would be to avoid being devoured by one of the predator species on the Martian surface, some of which they would be able to evade, and some of which were impossible to evade if encountered.

The Mars program was launched; Basiago and Stillings were told, to establish a defense regime protecting the Earth from threats from space and, by sending civilians, to establish a legal basis for the U.S. to assert a claim of territorial sovereignty over Mars. In furtherance of these goals and the expectation that human beings from Earth would begin visiting Mars in greater numbers, their mission was to acclimate Martian humanoids and animals to their presence or, as Major Dames stated during their training near Mt. Shasta in 1980: "Simply put, your task is to be seen and not eaten."

The most interesting whistle blower is Bernard Mendez, who claims that he attended jump room training classes taught by Major Edward Dames at the college of the Siskiyous in the summer of 1980. He said that other students at the class included, Andrew Basiago, future DARPA director, Rigena E. Dugan, Barak Obama, future astronaut William C. McCool and William Stillings. The 4 other students attending this class could not be remembered by himself or any of the others.

Both President Barak Obama and Edward Dames have denied their participation in the program. This could be because of their memories being erased, which was a typical procedure in these types of programs. Or, they could just be lying, perhaps for reasons of National Security. Or in Obama's case, the lying could be for political reasons. Admitting to the Mars Jump Rooms would seem like science fiction to the majority of voters and Obama's credibility could become worse than it already is.

Bernard Mendez claims that he was also investigating jump room technology for the U.S. intelligence community because of certain "bugs" in the system. The technology had been given to our government by the Grey ETs in a liaison program. The young participants were actually chosen by the Greys and then trained by the CIA.

There were two jump rooms, one in New York and one in El Segundo with a central control room in Ohio. Both Jump rooms were reporting 40 incidents of jump room participant injuries per month during teleportation to unknown environments in space. However the "injuries" seemed only temporary and would clear up in several days. Also sometimes there were power failures at the jump rooms while the central control room in Ohio indicated no problem at either site.

Mendez's job was to determine if the Greys had been secretly controlling the Jump Room technology for their own covert agenda. He was deployed, along with an evaluation team that included several prominent U.S. Astronauts and Barak Obama, to determine the causes of these problems and the true destination of these jumps rooms. While attending the class, Mendez debriefed the instructor Ed Dames on the intelligence situation.

After many jumps from the El Segundo jump room, they determined that sometimes they weren't going to Mars, but rather to a synthetic quantum environment (SQE) like folds in space time. The U.S. government has identified 153 SQEs constructed by the Greys in the near Earth environment, ranging from 400 miles above the Planet to the surface of the planet and under the surface of the planet. Since, these SQEs represent "new land" that might be territorially acquired, the U.S. Government is actively exploring them with a view of making them part of the United States territory.

Apparently the Greys selected Barak Obama in the 1980s and with the ability to time travel and see the future, he has been groomed to be president since that time. According to Richard Boyland, the same could be said for President George W. Bush who was actually shown that he would be President in a future viewing device, long before it actually happened.

Another interesting person that was an attempted recruit to join the Mars colony was Laura Magdalene Eisenhower, the great granddaughter of President Eisenhower. Her fiancé was part of the secret program and was trying to persuade her to join him and live on Mars. She felt that her mission was on Earth – not Mars and declined. He pressured her to the extent that she broke off the relationship with him.

A joint statement by Andrew D. Basiago and Laura Magdalene Eisenhower in 2010:

> "It is a positive thing for the human race to put survival colonies on other planets. Earth has been struck by many cataclysms in the past, and so we should protect the human genome by placing human settlements on other celestial bodies. Yet, when secrecy surrounding such projects tempts government to rob the free will of individuals, and excludes humanity from debating a subject that implicates the whole human future, and diverts the destiny of a planet to serve an off-planet agenda, the conscience of a free people requires that such projects be undertaken in the bright sunshine of public scrutiny, not within the dark corridors of the military-industrial complex."

Another Mars colony eyewitness Michael Relfe, former member of the U.S. Navy was recruited as a permanent member of the secret Mars colony in 1976. In 1976, he was sent to the Mars colony and spent 20 years as a permanent member of its security staff.

In 1996, Mr. Relfe had his 20 year memories of Mars service selectively erased with an Electronic Dissolution of Memory (EDOM) device. Then, he was time-traveled via teleportation and age-regressed 20 years, landing back at a U.S. military base in 1976. Mr. Relfe then served six more years in the U.S. military on Earth, with no memory of his 20 year service on Mars, before being honorably discharged in 1982.

Eventually his memory was restored through a long process of kinesiology. His very interesting story and the methods of kinesiology used to recover his memory is obtainable at: http://www.themarsrecords.com/

In a two-volume book authored by his wife, Stephanie Relfe, B.Sc., *The Mars Records*, Mr. Relfe describes the two types of individuals at the secret Mars colony:

"To clarify: Remember there are two kinds of people that I remember.

"1. People visiting Mars temporarily (politicians, etc.) – They travel to and from Mars by jump gate. They visit for a few weeks and return. They are not time traveled back. They are VIP's. They are OFF LIMITS!!

"2. Permanent staff – They spend 20 years' duty cycle. At the end of their duty cycle, they are age reversed and time shot back to their space-time origin point. They are sent back with memories blocked. They are sent back to complete their destiny on Earth." (Vol. 2, p. 204)

In an interview, Michael Relfe also discusses the presence and functions of Reptilian and Grey extraterrestrials at the secret Mars colony. I am not sure of the identity of the interviewer "EL"

"EL: What about the Reptilians?

Michael Relfe: Yes. They are racially related (Draconians, Reptilians, Grays.)

EL: Do any Grays and Reptilians live on the Mars Base?

Michael Relfe: Yes, some are stationed there. I remember the Grays as doctors or technicians. I believe the Reptilians stay camouflaged (cloaked) most of the time. They prefer to appear human because they are naturally fierce-looking." (Vol. 2, page 205)

Another interesting source of information comes from an anonymous person called Captain Kaye who was interviewed by Dr. Michael Salla of the famous Exopolitics (extraterrestrial politics) site; http://exopolitics.org .

Captain Kaye claims that he as a child he was secretly trained to be a "super soldier", along with 300 other children, 20% of which were girls, in a program called Operation Moon Shadow. Operation Moon Shadow was a USMC special section program operated jointly with an extraterrestrial group known as the Bronze ones.

The secrecy was maintained using time travel technology. He would be abducted at night from his bed, perhaps using teleportation technology. He would do his training then he would be sent back in time to just after he was abducted and returned to his bed so no one would realize that he was even gone. Also, it would seem like he only dreamed about his training in his sleep.

Later in 1987, he was officially transferred into a special division of the Marine Corps at age 17 and after completing his training, was sent to Lunar Operations Command, a base on the far side of the Moon. At this base, he had to sign a special contract. The contract stated that he would serve 20 years in the Earth Defense Force. At the end of his 20 year term of service, he would be age regressed 20 years, his memory of his service would be wiped out and he would be sent 20 years back in time to start a new life.

Apparently the Earth Defense Force is a UN "Unacknowledged Special Access Program". This defense force recruits people from many military services globally.

He agreed to the terms, although they couldn't tell him where he would be serving and what his job description would be for security reasons. He was told that when he arrived at his destination, these things would be revealed to him.

Then, they were loaded in a craft shaped like a triangle that was three stories high, 400 foot wing span and 150 feet long that could carry about 2,000 passengers. There were quite a few empty seats on his flight. When he arrived at his destination, after a flight of several hours he discovered that he was on Mars. The door opened on the craft and the people on board walked outside. He was surprised that the Martian air was breathable; he estimated that it was similar to being at the 10,000 foot level on Earth.

He was told to walk over to another building at the base where he would be filled in on his mission. There, he discovered that he was at a base called *Aries Prime* that was built by the Mars Colony Corporation. *Aries Prime* was the main human headquarters on Mars. His mission, as a Mars Defense Force Marine, would be to defend property belonging to the Mars Colony Corporation from civilized native beings on Mars. These native beings consisted of two basic types, the Reptilians and the Insectoids, who were constantly testing their defenses with military attacks. The Mars Defense Force was a subsidiary of the Earth Defense Force.

After his briefing, he was loaded into a Mars shuttle antigravity craft which he described as being about half the size of a bus and which flew at an average height of 20 feet above the ground, so as not to present a target to the natives. Finally, he arrived at the Vulcan settlement, Forward Station Zebra - Division 098. Nearby, he could see Division 097.

Here, he was placed in a room with the other newbies and given more briefings. He was shown survival suits that were required for longer term surface outings and powered body armor which would even up the odds in fighting the Reptilians and Insectoids who physically were quite strong.

Then, his group was introduced to the other men and women of Division 098. Oh yes, about one third of the personnel at Division 098 were women. There were 244 persons in each Division. He realized that these people would probably be his companions for the next 20 years so, he had better be on friendly terms.

For the next three years, he was involved with numerous battles with the natives of Mars defending Mars Colony Company assets which included mining operations. Then, a new woman arrived at Division 098. For the next year, Captain Kaye and the new woman hated each other and were always arguing. And then, they fell madly in love and ended up getting married. Often, they fought side by side in battle, and each knew that the other would be willing to die to save them.

Later, in a battle his wife was killed with a severe head wound. This event totally saddened and depressed Captain Kaye. But, he had to carry on.

On the seventeenth year of Captain Kaye's tour of duty, the Mars colony had negotiated a peace treaty with the natives. This was an important treaty. It meant that all the unnecessary warfare between the Earth colonialists and the natives of Mars could come to an end.

But, even before the ink was dry on the treaty, so to speak, the Mars Colony Corporation wanted four divisions of the Mars Defense Force to raid a certain Reptilian Temple to recover some artifacts there that they greatly desired. It would seem that corporate greed and basic lack of wisdom and integrity knows no bounds.

When the personnel at Division 098 heard the orders, they knew that it was a terrible idea. Captain Kaye went to the commander of the Division and voiced his concerns. The Commander agreed with him but said he had to follow orders from the "brass upstairs".

So, a force of 976 attacked the Reptilian underground temple, fighting their way through the underground tunnels that led to the central area. When they got to the central area, which was about the size of the Astrodome, all their electronic communication equipment suddenly stopped working. They knew that was a bad sign.

Suddenly, areas that had seemed like rock opened up and swarms of Reptilian fighters poured through the openings and surrounded them. The slaughter was immense. Finally a communications person was able to get through to central command using a scalar communicator. Central command was able to project a worm hole over the battle area and extract the fighters. People unfortunate to be on the worm hole event horizon were sliced apart.

Of the original 976 Mars Defense Force soldiers only 35 survived. It was a total disaster! The personnel in those four divisions could not be replaced and Forward Station Zebra was shut down. Luckily though, the Treaty stayed in place. The Reptilians figured that, since the Mars Colony Corporation lost nearly four divisions of their best fighters, they had been taught a lesson and wouldn't repeat the same dumb mistake in the future. And after that, each side pretty much respected the boundaries set by the peace treaty.

After Captain Kaye was released from the hospital, he was asked if he wanted to attend flight school. He readily accepted and attended flight school for three months at Lunar Operation Command on the Moon and then three months at another base on Titan.

After flight school, he was stationed on the Intergalactic High Liner series mother ship named the Star Ship Nautilus (SS Nautilus) which acted as a space going air craft carrier. Captain Kaye estimated the SS Nautilus to be about one half to three quarters of a mile long. The forward section was the command area where flight navigation, diplomatic meetings and strategy meetings took place. The ship's officers also lived in this section. The midsection was the fighter, bomber, weapons, and supplies storage area and the flight crew living quarters and the rear section was the propulsion area. The propulsion was a combination of antigravity and temporal drives.

The 06 Captain of the ship, which displayed "Blue and Gold 213" on a plaque in the captain's office, was Roger L. Kirkland. Once, Captain Kaye got to meet him personally when he got "chewed out" for disobeying a rule dealing with female members of the crew.

Fighter craft consisted of three different classes; Vipers, a one man fighter, Cobras a two man (pilot and gunner) fighter and Rattle Snakes, which had advanced inter-dimensional sensors and often would require a three man crew, (Pilot, gunner and sensor systems operator) all used antigravity propulsion. The missions on these craft would be limited to 24 hours because of the danger of space madness to the pilots. Space madness occurs because of the lack of any ground references in deep space. Otherwise, the craft could have duration of 30 days.

These craft were loaded up with weapons and supplies inside the mother ship and then sent to the top with elevators and launched from the top through air locks. When Captain Kay asked where the SS Nautilus was built, he was told that that information was above his pay grade. Information was compartmentalized and no contact with Earth was allowed for the crew.

Captain Kaye didn't know how many other ships like the SS Nautilus were in the Earth Defense Force fleet, but at one time, he saw five of this class of mother ships close together. Although this class of mother ship had interstellar and intergalactic capability, while Captain Kaye was onboard the SS Nautilus, operations were confined to this solar system.

Captain Kaye flew all three of these type fighters but only flew one bomber class, the Albatross class bomber, with a delta shaped wing span of eighty feet. On one bomber mission, a Zeta base on Ganymede was bombed.

After three more years of service on the SS Nautilus, Captain Kaye and a few other men were told their time of service was finished and were sent back to Lunar Operations Command on shuttle craft that was smaller than the one that brought him to Mars.

There was an "end of service" banquet for all the men that had finished their tour of service. The Year was 2007. There were three speakers that gave speeches that said how much their service had helped to defend Earth and how much they were appreciated for the many sacrifices they had made. Captain Kaye didn't know who the first two speakers were. He had been out of contact with Earth for 20 years as they weren't allowed any contact with Earth. However he recognized the last speaker. It was Donald Rumsfeld!

Before he had his memory wiped, he thought of how much he had loved his wife and didn't want to forget her. Then, he was back on Earth in 1987, looking the same age as when he joined the marines as if nothing had happened at all.

However, one problem was that he was experiencing symptoms of Post-Traumatic Stress Disorder. Another problem was that he was actually existing on two different time lines simultaneously.

Both problems led to psychological problems with Captain Kaye. He sought help from psychiatrists and gurus and started meditating in order quiet his rational mind and to access his sub-conscious mind.

Eventually, memories of his wife would start coming through. Then, more memories would come back. At first, he thought that he was crazy. But, after many years, the whole picture finally presented itself to Captain Kaye by age 37, when his other time line came to an end.

His story, which Exopolitics is trying to verify with other documentation, is in some ways similar to Michael Relfe's story. Michael Relfe also did a twenty year tour of duty on Mars, was age regressed, had his memory of twenty year service wiped, and time traveled back to the time of his enlistment in the Navy.

Later, after Michael Salla's original interview with Captain Kaye, new information was received. The mysterious Captain Kaye revealed his true name as Randy Cramer.

When Randy Cramer was trying to obtain certain documents through his Congressman, his former Commander, USMC Colonial Jamieson contacted him and put him in touch with a man he didn't know, who claimed to be a Brigadier General Smythe, who said his orders had changed and now Randy Cramer was to go public with all that he knew, including the names of these officers.

Now, Randy Cramer was simply following orders to go public from his direct chain of command. Apparently there are groups within the Earth Defense Force that want more public revelation of what is really going on.

So, if Andrew Basiago is making all this teleportation, a base on Mars and time travel up, he has a lot of company. Furthermore,

Andrew Basiago claims that his involvement in Project Pegasus was confirmed by the White House in a secret meeting held at Wolf Creek Pass Ski Lodge in Southwest Colorado in June 2003.

Another startling revelation by Andrew Basiago is that he could see past and future historical events on a device called a "cronovisor" and that he actually viewed the 9/11 attacks of 2001 via cronovisor decades before the event. So, certain government officials knew about the 9/11 attacks decades beforehand.

On the other hand, Basiago stated that he would release a book about his experiences in 2011 titled: *Once Upon a Time in the Time Stream: My Adventures in Project Pegasus at the Dawn of the Time-Space Age*, however to this date (2014) it still is not available. Maybe the secret government got to him. Maybe it exists on another time line. I really don't know why it is not out there.

There exist many different people, many in distinguished professions that claim there is a Mars Base that they have been to. And some have stated that they were sent there with teleportation technology partly developed from Tesla technology and the Philadelphia Experiment and partly given to MJ-12 by the Greys, dubbed "Jump Rooms."

So there exists technology that goes way beyond antigravity and allows for travel anywhere in space and time. If this concept is too difficult to wrap your mind around, for your own mental comfort you could consider this book to be a science fiction novel. But if you do the research, as I have done, you might realize that much of it could actually be true.

10

Conclusions

There are still attempts by certain parties to debunk the historical existence of MJ-12 and the 1947 Roswell crashed UFOs incident. However, the overwhelming evidence states otherwise.

For example, Jesse Marcel had brought some of the crashed UFO debris home, which was seen by his 11 year old son, Jesse Marcel Jr. Strange I-beams from the wreckage had violet symbols inscribed on them. Jesse Marcel painstakingly copied these symbols on paper and saved this paper after returning all the debris, as ordered, to the Roswell Army Air Force base.

Sixty years later, after no longer being in the military, His son, Jesse Marcel Jr., wrote *The Roswell Legacy* which reproduces these symbols exactly. Moreover, these symbols have been translated and that translation is also in the book.

Timothy Good in his book, *Need To Know,* has provided plenty of information about the Roswell UFO crash and recovery. And Phillip J. Corso, a former intelligence officer who served on President Eisenhower's NSC staff and headed the Army staff's Research and Development Foreign Technology Desk at the Pentagon, wrote *The Day After Roswell,* provided even more information. So, there are a number of distinguished researchers and whistleblowers that have come up with pretty tangible evidence that the Roswell crashed UFO incident did indeed happen.

As far as the Eisenhower Briefing Documents go, Stanton Friedman spent over a decade verifying their authenticity in great detail. Similar care was given other revealed MJ-12 documents. Most sincere UFO researchers are convinced of their authenticity. The debunkers have failed to prove otherwise. The consensus is that these matters can be considered as historical fact even though there are some powerful organizations that wish this history to remain secret.

An organization that has investigated the secret space program has recently had a conference (June 28-29, 2014) in San Mateo, California with many speakers presenting their information. Their website is here: http://secretspaceprogram.org

Most of the other information in this book will eventually also be considered historical fact in the future. Although I admit that some of the information may not pass the test of close scrutiny as done by researchers like Stanton Friedman.

So apparently, the United States and other countries in conjunction with the United Nations and the extraterrestrial Star Nations are operating a secret program called Solar Warden to police the space of our solar system. The technological machinery to carry this out has been under development since before World War II, even before the twentieth century, if one included Keely's and Tesla's work.

The fallout from World War II created the political climate to bring into existence all the national and international organizations to carry out this secret project, like the UN the NSA and the

CIA. The following "Cold War" between the two former allies was, mostly likely, a smoke screen to create fear in the public of Russia and the U.S. so that they would willingly go along with vast military expenditures and the "Space Race" while, secretly, both countries were cooperating in a secret space program.

After World War II, the technology developed by the Germans, some of which was derived from extraterrestrial sources, was to a great extent taken over by the U.S. and other allies. To this was added UFO crash retrievals by the U.S., from which back engineered technology was also developed.

Also, secret, but unlawful, treaties were made between the U.S. and certain alien civilizations which wanted to do genetic testing and experimenting on abducted humans to further their own chance of biological survival. The U.S. agreed to their wishes on conditional terms in return for their technology and instructions on how to reproduce and use it.

The leaders of the U.S. felt that they had to do this to bide their time until they could develop their weapon technology sufficiently to defend themselves from attacks by extraterrestrials with technology, that at the time, was far in advance of our own.

So, by the early 1960s, the U.S. had its own operational anti-gravity flying saucers and other types of space craft which were developed in secret "black projects" largely financed through Waivered Unacknowledged Special Access Projects with money diverted from other military programs. By 1962, the Russians and the U.S. jointly landed on Mars and by 1964, they had established a base there. By the late 1960s, the CIA was also operating time travel and teleportation technology in Operation Pegasus.

Because of extended secret negotiations between various Star civilizations which had formed an alliance called the Star Nations, and the U.S., Russia, China and the precursor to the European Union, the European Council, by 1964, a secret treaty had been arrived upon between these groups.

This treaty however was violated numerous times by a certain faction called the Cabal (or New World Order) because they feared losing power over the people of the Earth when full disclosure of the treaty with the Star Nations would be revealed and the people of the world discovered what was really going on. The Cabal's goal was to sabotage the treaty.

The Cabal violations of the treaty became so numerous and blatant that the Star Nations declared that the treaty was null and void.

However other groups saw the need for a treaty and continued negotiating. By the 1980s, the situation was resolved and placed back on track. And eventually, the secret treaty came under the management of the U.N. so no one world power would gain total control.

Secret diplomacy between Earth governments and Extraterrestrial ones suffer from a basic flaw. Secret Diplomacy fails to represent the will of the people of Earth and only expresses the will of an extremely small, select portion of Earth's society.

The Solar Warden program is supposed to protect our solar system area of space from the "Bad Guys". I certainly am not opposed to the ideas behind this program. What I am opposed to is all the secrecy involved. To preserve all this secrecy, an unfair, unnecessary and tremendous strain is placed upon the service men and women of the Earth Defense Force. Also, a lot of money has been wasted on rockets to get into space because of the desire to keep antigravity technology secret.

If the Solar Warden space program is so necessary, I am certain the national and world public would approve. So, why not remove the secrecy?

The purpose of this book is to cast some light into the shadows of secrecy and to uplift science and technology to a higher level. Why keep teaching obsolete science to our young people? After all, they represent our future.

Would you prefer a future world based on ignorance, secrecy and the mistrust that secrecy engenders or on enlightenment and openness? The correct answer is fairly obvious.

Why keep denying that our governments have been in contact with extraterrestrials for some time? The governments should just come clean and admit what is going on. Is it because there is a secret Military Industrial Extraterrestrial Complex cabal that wishes to keep all this advanced technology to themselves for their own selfish reasons? The greedy history of the Rothschilds, the Rockefeller family, the Vatican and their ilk certainly has shown this to be the likely scenario.

Some technical secrecy is probably a good idea in areas where the technology could be misused. But then, we already have dangerous technology. How many people are killed in car accidents every year? What about nuclear, biological, and chemical weapons? Secret and not so secret agencies are already abusing these types of weapons.

Even after Chernobyl and Fukishima, the nuclear industry is still trying to sell dangerous reactors. These reactors breed weapons grade plutonium and even more radioactive waste material. The radiation is a health hazard. This type of technology should be outlawed for the sake of preserving an ecologically healthy planet.

When I was young and idealistic, I was passing out anti-Vietnam War leaflets. An owner of a gas station informed me that I shouldn't be concerned about the Vietnam war because war was good for the economy. Well actually, war is not good for the economy. It only temporarily seems that way because all the government spending on the war effort stimulates the economy. In the long run, war is inflationary because a lot of money is spent with little real assets being created. In war, real assets are being destroyed in someone else's country. If the war was fought in our country, we would quickly realize why war is actually bad for the economy.

The government is the largest employer and most sensible people realize government spending is necessary to keep our economy going. In the future, some means of wealth distribution will be required as robots replace humans as the primary producers of goods. Corporations prefer robots that can work 24/7 with no vacation or sick leave. As in a game of monopoly, the economic game ends when one person has all the money. Unemployed consumers that have no money will purchase no products. Hence, there will be no economy. End of game.

But, why not get real value for government spending instead of spending on destructive war? If the government spent that much money on a creative project rather than a destructive one, our nation would be much better off, as would be the world.

Why not spend the money that we waste on un-necessary war and destroying other people's countries on a colonizing of space project? The companies now making "black project" weapons could apply their high tech for a positive, non-destructive purpose by producing space colonizing hardware. We are worried about over population on this planet. One solution is to move some people off planet to colonize space.

Space colonization could include colonization of habitable planets and moons of this solar system and less habitable locations with the appropriate climate and atmospheric enhanced infrastructures and possible terraforming projects. Mining the asteroids could also yield economic rewards that could help finance these projects. Existing antigravity and teleportation technology makes it all economically feasible. It just requires it to be made open, rather than "black" technology.

This colonization should not be in secret projects like the Mars Colony Corporation, and the secret space program, but in open projects that would be less expensive because the high costs of secrecy are no longer needed. The adventurous people who volunteer for these projects would then be given the recognition they deserve for their efforts instead of being un-recognized as they are now in these ultra-secret projects. This could be a *good* Plan for a New American Century, in place of the highly destructive and illegal present PNAC plan, which is also destroying the reputation of the United States government among the world's people while lessening our freedom at home.

The powers that be have been keeping all this secrecy for much too long. Fear, hatred, greed and vengeance are all based on spiritual darkness and ignorance of higher realities. As the ethical ETs have warned, our present course is leading to planetary destruction. It is time to leave all that spiritual darkness and stupid foolishness behind and for our civilization to straighten up, evolve to the next step and become a member of the galactic civilization with all the tremendous benefits that would entail.

END

Biblography

1. *Universal Laws Never Before Revealed: Keely's Secrets: Understanding and Using the Science of Sympathetic Vibration* by Dale Pond.

2. *Nicola Tesla Colorado Springs Notes 1899-1900* by The Nicola Tesla Museum, Beograd, Yugoslavia.

3. http://www.proliberty.com/observer/20070405.htm and *The Bush Connection* by Eric Berman

4. *The Energy Revolution* by Callum Coates.

5. *Briton's Secret War in Antarctica* Nexus Magazine Volume 12, Number 5 and http://www.bibliotecapleyades.net/tierra_hueca/esp_tierra_hueca_13.htm

6. *The United States and Germany's UFOs from 1917 to the Present Day* by Maximillien de Lafayette

7. *Kennedy's Last Stand: Eisenhower, UFOs, MJ-12 & JFK's Assassination* by Michael E. Salla, Ph.D.

8. *The Philadelphia Experiment Murder* by Alexandra Bruce, Edited by Peter Moon.

9. *Exposing U.S. Government Policies on Extraterrestrial Life: The Challenge of Exopolitics* by Michael E. Salla, Ph.D.

10. *Extra-Terrestrials Friends and Foes* by George Andrews.

11. *http://www.drboylan.com/mj12unoosacss.html*